THE POPULATION-SAMPLE DECOMPOSITION METHOD

INTERNATIONAL STUDIES IN ECONOMICS AND ECONOMETRICS

Volume 19

The Population-Sample Decomposition Method

A Distribution-Free Estimation Technique for Minimum Distance Parameters

by

A.M. Wesselman

1987 **KLUWER ACADEMIC PUBLISHERS**
DORDRECHT / BOSTON / LANCASTER

Distributors

for the United States and Canada: Kluwer Academic Publishers, P.O. Box 358, Accord Station, Hingham, MA 02018-0358, USA
for the UK and Ireland: Kluwer Academic Publishers, MTP Press Limited, Falcon House, Queen Square, Lancaster LA1 1RN, UK
for all other countries: Kluwer Academic Publishers Group, Distribution Center, P.O. Box 322, 3300 AH Dordrecht, The Netherlands

Library of Congress Cataloging in Publication Data

```
Wesselman, A. M., 1956-
  The population-sample decomposition method.

  (International studies in economics and
econometrics ; 19)
  Bibliography: p.
  Includes indexes.
  1. Multivariate analysis.  2. Sampling (Statistics)
3. Estimation theory.  I. Title.  II. Series:
International studies in economics and econometrics ;
v. 19.
QA278.W47  1987      519.5'35      87-21446
```


ISBN-13: 978-94-010-8147-4   e-ISBN-13: 978-94-009-3679-9
DOI: 10.1007/978-94-009-3679-9

Copyright

Acknowledgements

This volume has been written in a time span of four years. From May 1983 until January 1985 I was affiliated to the Centre for Research in Public Economics of the Leyden University and during the remaining period (until May 1987) I was affiliated to the Econometric Institute of the Erasmus University in Rotterdam. The research has been made possible by financial support from the Netherlands Organization for Advancement of Pure Research (ZWO).

I am grateful to my colleagues, who worked with me during thesé four "PSD years"; especially the "Leyden-group", that moved with me from Leyden to Rotterdam. With their advices they contributed directly or indirectly to the final results of the research.

I am especially indebted to Bernard van Praag, who initialized the PSD-approach and encouraged and advised me in many stages of the project. Finally I would like to thank Ellen Biesheuvel and Aagske Sporry at the Erasmus University for their careful typing of the manuscript.

THE POPULATION-SAMPLE DECOMPOSITION METHOD:
A DISTRIBUTION-FREE ESTIMATION TECHNIQUE FOR MINIMUM DISTANCE PARAMETERS

Table of contents

I. Introduction to the Population-Sample Decomposition Approach

I.1. The linear statistical model.

One of the basic objectives of the social sciences is the drive to discover
regularities within a set of objects, usually called the population. If we
were able to obtain a complete set of information about those phenomena which
are of interest to us, we would be in a position to make deterministic
statements. However, as the population under consideration is often too large
to be observed in its entirety, it is impossible to obtain the necessary
information from all the objects of the population. It is for this reason that
the phenomena are only observed for a subset of objects, that are randomly
sampled from the population. Let us assume that the resulting set of
observations can be described by a set of vectors. The observed vectors are
collected in a dataset, called the sample. The assumed regularities in the
population must be a reflection of corresponding relations in the sample
observations, as the sample represents the phenomenon to be observed. In this
way, the real-world regularities can be estimated by means of the sample
observations.

The aim of this study is to provide a solution to the problem of how to use
sample information subject to minimal assumptions with respect to linear
statistical models, that are of interest in multivariate statistical analysis.
By using the word "model" a direct reference is made to a second aspect that,
together with the set of sample observations, plays an important role in
statistical theory. In multivariate statistical analysis one starts by
creating a model on the basis of intuitions about relationships in the
population. The model is assumed to be an approximation of reality and gives a
formal description of the data generating process. The parameters of the model
are estimated by means of the observed sample values and the distributional
properties of the resulting estimates are based on the assumption that the
data are generated by this particular model. The question arises as to whether
it is always permissible to postulate that the hypothesized model is a good
approximation of reality. Naturally, if some hypotheses are implied by the
theory, or by empirical evidence from previous research results, then they
should be included in the model. However, one often finds that a part of the
model forms a dubious approximation of reality and is merely postulated for

the sake of convenient estimation. Before we continue with an outline of an alternative approach to the estimation of statistical relations based on minimal model assumptions, attention will be paid to some general ideas behind data analysis and modeling in multivariate statistical analysis. Firstly, some ideas about exploratory data analysis, based on non-random sample values, will be formalized (see also Tukey (1977)). Secondly, these ideas will be extended into a propabilistic context, yielding confirmatory data analysis based on model assumptions and probability theory, and statistical inference techniques for the resulting estimators.

Suppose that $\{x_n: n=1,...,N\}$ is a set of N observations, the values of which are indicated by the (k×1) vector x_n. That is, x_n stands for the value of the nth observation on the vector of interest. If the relevant aspects of the population are reflected by a (k×1) vector x, then a functional specification of the regularities in the population can be formalized as

(1.1) $f(x;b) = 0$,

with f being a function that is assumed to be known up to its parameter vector b. The problem is how to derive the value of b, by means of the N values of x given in the sample. The hypothesized functional form of f constitutes, when relation (1.1) is placed in a probabilistic context, a part of the statistical model. However, in order to start with a data descriptive technique, we do not yet assume the existence of any randomness. Consequently, relation (1.1) should apply to each observation x_n, so that we may write

(1.2) $f(x_n;b) = 0$ for n=1,...,N.

However, generally it is not possible to find a suitable value for b, and we hall have to make do with a value for b in the sense that (1.2) will hold "as ccurately as possible" for all observations, according to some criterion of fit. If δ_n is defined as the deviation of $f(x_n;b)$ from zero, hence

(1.3) $\delta_n = f(x_n;b)$ for n=1,...,N

then a well-known criterion of fit is given by the sum of squares

(1.4)
$$\sum_{n=1}^{N} (\delta_n)^2.$$

Minimizing (1.4) with respect to b, an optimal value of b, say $\hat{b}$, is obtained as a function of the sample values $x_1, \ldots, x_N$ and $\hat{\delta}_n = f(x_n; \hat{b})$ can be seen as the deviation of x_n from the hypothesized functional approximation.

So far, no randomness has been assumed and nothing has been said about the reliability of the value $\hat{b}$. The vector $\hat{b}$ is merely a descriptive statistic. However, in the early thirties, the realization dawned that the resulting value $\hat{b}$ varies, when it is calculated on distinct samples that are dealing with the same phenomenon. Especially under the influence of R.A. Fisher (much of modern statistical methodology is founded on Fisher's classics (1925, 1935)) these methods have been embedded in a probabilistic context. That is, the sample can be seen as a realization of a stochastic event and the resulting estimator $\hat{b}$ should be considered as a realization of a stochastic event too. So now we are interested in the statistical properties of the resulting estimator. In multivariate statistical analysis this problem is solved by assuming a specific data-generating model, which is specified up to a number of unknown parameters. These parameters are estimated from the observed sample and inferences are based on the assumptions about the stochastic structure of the model. A way of introducing the stochastic concept into the previously described data analysis, is by assuming that one or more of the components of x_n are random in character. One may distinguish the random variables from the fixed values by partitioning x_n into subvectors Y_n and z_n, where the random subvector Y_n is denoted by a capital in order to emphasize its stochastic character. As a consequence, the function $f(Y_n, z_n; b)$ is random as well and (1.3) is rewritten as

(1.5) $f(Y_n, z_n; b) = \varepsilon_n$ for $n = 1, \ldots, N$

with $\varepsilon_1, \ldots, \varepsilon_N$ a sequence of random error terms, that are usually assumed to have zero expectations. Assumptions about the functional form of f and about the distribution of ε_n are called model specifications.

It is worth noting that the data are supposed to comply with the model. Consequently the estimation methods and the statistical inference about the estimators are based on these specific assumptions. In general, stronger model assumptions yield more precise estimators. However, one should realize that these model assumptions should always describe the real-world situation. Although more restrictive model assumptions may suggest an increase of the statistical reliability of the estimators, this may turn out to be an artefact, and besides they may also increase the level of bias due to misspecification. If model assumptions are invalid, the result is that statistical inference does not give any information about reality. More about the choice of models, based on probabilistic theory can be found in McGullagh and Nelder (1983) and De Leeuw (1984, 1987). The advantages and disadvantages of using models in multivariate statistical analysis have been discussed by Box (1979), Tukey (1980), Kalman (1982,a) and Gifi (1984). They emphasize the importance of a good model, because as in cases of misspecification, the misspecification errors may exceed the sampling errors.

In this study we shall concentrate ourselves on linear relations. This means that function f is a linear function described by

$$(1.6) \qquad b'x_n = \delta_n \qquad \qquad \text{for } n=1,\ldots,N$$

where it is assumed, without loss of generality, that the observations are taken in deviation of the sample mean. Placing this relation in a probabilistic context, one may derive the statistical properties of the parameter estimators subject to the assumption that the regularities in the population are supposed to be indicated by a linear combination of the components of x, so that (1.6) approximates reality. The origin of these linear relations is geometric. The vector observations $x_1,\ldots x_N$ are interpreted as N points in a k-dimensional space. The intuitive idea behind the least-squares procedure is to determine in R^k a (k-1)-dimensional hyperplane, in such a way that the observations are "as close as possible" to that hyperplane See also Pearson (1901), who gave a solution of the least-squares method, in the case of a linear relation as described in (1.6).

The least-squares method has a long history. It has already been introduced in the context of linear relations at the beginning of the 19th century by Legendre and Gauss (see Plackett (1972)). They derived an optimal value for the vector b in (1.6), when it is assumed that the first component of b is fixed and equals -1. This method laid the foundation of the well-known linear regression model, in which a (random) variable Y_n is explained by a linear combination of predetermined values, that are denoted by a vector z_n. This linear regression model can be considered as an extension of the data descriptive technique given in (1.6), and is arrived at by partitioning x_n into a random variable Y_n and a $((k-1) \times 1)$ vector z_n of fixed values. Partitioning vector b accordingly as $b' = (-1, b'_z)$ and introducing a random character for the error terms, relation (1.6) transforms into

$$(1.7) \qquad Y_n = b'_z z_n + \varepsilon_n \qquad\qquad \text{for } n=1,\dots,N.$$

With the additional assumption that the errors $\varepsilon_1, \dots, \varepsilon_N$ are independent random variables that have zero mean and identical variances, we arrive at a description of the linear regression model in its most elementary form. An estimator of b_z can be obtained by minimizing the squared errors and the statistical inference is based on the assumptions about the distribution of ε_n.

In order to illustrate some modeling concepts, an example will be considered that is widely used in econometrics, viz. the simultaneous equations system. As the author of this study is most involved in the econometric branch of statistical applications he is most familiar with the application of statistical theory and modeling concepts in economic theory. This is the reason that, not only here, but also in the chapters to come, most of the discussions of statistical theory and references to other works have an econometric tendency.

The simultaneous equations system, the foundations of which were laid by Haavelmo (1943), forms one of the most important models applied in econometrics. In this model one assumes a dichotomy in the variables. That is, the observation vectors are partitioned into two distinct sets of variables; the variables of the first subset are treated as fixed, non-random values (the so-called independent or exogenous variables) and the variables of the second subset are assumed to have a stochastic character (the dependent or endogenous variables). It is hypothesized that the values of the endogenous variables are determined through the joint interaction with other variables, yielding causal linear relations. The values of the exogenous variables, that affect the outcome of the endogenous variables, are determined outside the system. Within

the block of endogenous variables it is impossible to distinguish between
causes and effects, because of the two-way dependency which holds between
these variables. When considering the disturbance terms in the model
definition, it is common practice to assume a (normal) distribution. Usually
all the model assumptions, that concern the identification and estimation of
econometric relations, are based on economic theory and empirical evidence
from previous research results. The distributional assumptions form the basis
for the statistical properties of the parameter estimators and are, for
instance, used to determine statistical tests of hypotheses on the values of
the parameters, and to obtain predictions of variables which have not yet been
observed. (Extensive reviews of econometric theory and introductions to
simultaneous equations systems can be found in various econometric textbooks,
like Goldberger (1964), Malinvaud (1970), Theil (1971), Maddala (1979), Judge
et al. (1980)). The statistical properties of the estimators and the outcome
of tests do not only depend on the sample information, but are also greatly
influenced by the model assumptions. Most econometricians are fully aware of
the fact that if one ignores model misspecifications, this may result in
invalid statistical inference about the parameter estimators and in many
cases, misleading claims of significance. Therefore a lot of attention has
been devoted to the development of procedures to test the adequacy of the
model to describe real-world phenomena; see for instance Ramsey (1969), Byron
(1972), Pesaran (1974), Hausman (1978) and Amemiya (1980). Kiviet's monograph
(1987) focuses on various aspects of econometric model misspecification tests
and provides an exhaustive review of the recent literature on the subject.
White (1981, 1982) does not only present several misspecification tests, but
also examines the consequences of maximum likelihood estimation of
misspecified models. A collection of papers, that covers a wide range of
modeling problems is to be found in Dijkstra (1984).

The idea of dealing with economic modeling in a probabilistic context
originates from Haavelmo's work (1944) on the statistical approach in
econometrics. However, the problems that arise from the statistical modeling
concept have led to various proposals for alternative methods to approximate
population regularities. For instance Kalman (1983) states that econometric
theory should follow a completely different route. He claims that econometrics
is too much dominated by statistics, whilst at the same time, probability
theory has nothing to say about modeling concepts in economic theory. His

reasoning stems from the idea that the statistical exercises are often based on "prejudice" rather than on reliable model assumptions that describe reality. As an alternative he proposes that the study of econometrics be placed in a system theoretic context (see also Kalman (1982,b)). On the other hand, Deville and Malinvaud (1983) confirm the importance of the inferential point of view in econometrics. Regression analysis should not be considered as the only method for dealing with multidimensional observations. The French school of economic statisticians have recognized the usefulness of other techniques, that are of a purely descriptive type, in order to reduce the dimensionality of a dataset. They claim that, for instance, Principal Components analysis tends to provide an appropriate alternative approach for exploratory economic-data analysis (see Benzecri et al. (1973) and Lebart, Morineau and Tabard (1977)).

In this study the statistical properties of well-known multivariate statistics will be reconsidered subject to alternative distributional assumptions. The starting point of our analysis will be a random sample that reflects the real-world phenomenon of interest. That is, in contrast to the assumption of fixed predetermined sample values, like those hypothesized for the exogenous variables in simultaneous equations systems, all the variables of the observations are assumed to be stochastic in character. The reason for doing so is that sample observations, in many cases in the social sciences, can not be treated as fixed non-random variables. This will not be the case when the exogenous variables are controlled by the researcher, like for instance in the case of a laboratory experiment. However, the case of controlled experiments is an exception rather than the rule in statistical experiments in social sciences.

The fact that we shall not distinguish between random variables and fixed values, is the reason for the random sample assumption. In order to emphasize the stochastic character of the sample, the observation vectors will be denoted by capital X's, so that from now on the random sample is given by $\{X_n: n=1,\ldots,N\}$. Kendall and Stuart (1961) and Kendall (1951, 1952) also distinguish between models that include fixed valued variables and models that deal with random variables only, which they referred to as functional and structural linear relations respectively. In the context of structural linear relations, Robertson (1974) and Mak (1983) may be mentioned, who derived the maximum likelihood estimators of regression vectors, given that the dependent and the independent variables are simultaneously normally distributed. The structural versions of linear statistical models are analysed in this study by means of the Populations-Sample Decomposition (PSD) approach; a method that has been discussed before in Van Praag, De Leeuw, Kloek (1986). Linear model parameter estimators are derived by means of projection procedures and the asymptotic properties of these estimators are obtained without assuming any specific distribution for the observation vectors. A similar distribution-free estimation technique has been applied in the context of the linear regression model in Van Praag (1980). (More references to linear regression analysis with random regressors can be found in section III.4, that discusses the regression case in the PSD approach). This study will illustrate how the PSD method can be applied to a much wider class of statistics. The large sample properties of multivariate statistics like principal factors, canonical correlations, principal components and some goodness-of-fit measures will be obtained subject to minor model assumptions.

I.2 Minimum distance parameters subject to minimal model assumptions

The main purpose of this study is to weaken the model assumptions of classical statistical models in two ways. Firstly, linear statistical relations are fitted to the data, without assuming that this relation reflects the data generating process. The linear relation merely describes the subspace that "fits the data as accurately as possible". The parameters will be introduced within a context similar to that of a structural linear model. That is, it is assumed that the sample observations $\{X_n: n=1,\ldots, N\}$ are random vectors, all distributed like the phenomenon of interest, that will be denoted by a random (k×1) vector X.

The second relaxation of the model specification is that we make minimal assumptions about the distribution of X in the sense that it is only assumed that the moments of X, used to describe the model parameters and the statistical properties of their estimators, exist. In this study we shall find that this results in the assumption of finite second-order moments, for the definition of parameter estimators. The existence of the fourth-order moments warrants the application of limiting theorems, yielding asymptotic statistical properties of the estimators and especially finite variances. It might be possible to drop the finiteness of fourth-order moments, but that would imply that our estimators, that are functions of the second-order sample moments, do not have a finite covariance matrix. If the requirement that second-order moments exist is also dropped, this would make the estimators themselves meaningless. So, our minimal assumptions are minimal in the sense that without them, no sensible large-sample theory about estimators involving second-order moments can be carried out.

In this study we shall assume throughout that the observations are independent and identically distributed (i.i.d.), in order to show the convergence in distribution of sums of random vectors by means of the Lindeberg-Lévy central limit theorem. For this purpose alternative central limit theorems may be used, which are warranted in cases of less restrictive assumptions than the i.i.d. assumption. It follows that a lot of this theory can be extended to non-i.i.d. observations. However, these complications are left outside the scope of the present study. The main difference from the classical model assumptions is that we do not postulate a functional form of the data generating process. That is, instead of adopting a linear relation that should apply to the observations, we work the other way round. Starting with a set of observations, a linear relation is fitted, without assuming this linear relation to be true for this particular dataset.

In order to explain what is meant by "a subspace that fits the data as accurately as possible", some geometric concepts are introduced. Let us assume that we have a set of non-random variables, the values of which are given by $\{x_n: n=1,\ldots,N\}$, where the k-vectors $x_1,\ldots,x_N$ can be represented by N points in a k-dimensional space. In (1.2) we have defined a (k-1)-dimensional linear subspace by minimizing the sum of squared distances between the N points and this subspace. With respect to this particular dataset, the subspace, say S, is in accordance with the minimum squared distance criterion in an optimal position. When a linear relation really exists between the components of the vectors $x_1,\ldots,x_N$, then the N points are scattered in at the most (k-1) dimensions and the minimum sum of squared distances will be zero. In general the minimum sum of squared distances will be greater than zero, since the N points are not situated in one (k-1)-dimensional subspace. The fitted linear relation is fulfilled exactly by the projections of the observations in S and it forms an approximation of a linear relation in the observations themselves. Similarly one may project the N points onto a (k-p)-dimensional ($1 \leq p < k$) linear subspace S. Denoting the projections of x_n by $\tilde{x}_n$, the optimal (k-p)-dimensional projection space can be described by p linear relations in $\tilde{x}_n$ as

(2.1) $B'\tilde{x}_n = 0$ for $n = 1,\ldots,N$

with B a (k×p) matrix of rank p. By extending this geometric insight into a probabilistic context, one may get a good geometric interpretation of many linear statistical methods. Actually, most multivariate methods in statistics have a geometric origin (see for instance Anderson (1984)). It is the geometric interpretation of linear statistical models that forms the foundation of the minimum distance approach.

The exercises in relation to (2.1.) are placed in a probabilistic context by assuming a random character for the vectors $x_1,\ldots,x_N$. That is, these vectors are assumed to be random observations of an underlying random vector X, that represents the relevant aspects of the phenomenon of interest. By means of the minimum distance technique, we shall try to specify a vector $\tilde{X}$, say, that is "as close as possible" to X and for which the approximating relation holds

exactly. As $\tilde{X}$ is the projection of random vector X onto a (k-p)-dimensional
subspace S, this is a random vector as well and the difference X-$\tilde{X}$ is called
the residual vector. More explicitly, when it is assumed that vector X has
zero expectation (E(X) = 0) and that it has to satisfy approximately p linear
relations of the type B'X = 0 defining a subspace S, then these relations hold
for the projection $\tilde{X}$ of X in S, exactly. Hence

(2.2) $B'\tilde{X} = 0$

where B is a (k×p) parameter matrix, while in general B'X $\neq$ 0. In order to be
more specific about the statement "as close as possible", a distance measure
will be defined on the k-dimensional space R^k. $\tilde{X}$ is obtained as the k-vector
that is situated in the linear subspace described by (2.2) and that minimizes
the mean squared distance to X. In this study we shall restrict the distance
function to the Euclidean type, i.e. when x and y are two vectors in R^k, then
the squared distance between x and y is given by $d^2(x,y) = (x-y)'Q(x-y)$ with Q
a positive definite, symmetric matrix. As a consequence the projection $\tilde{X}$ is
obtained as a linear function of X and the mean squared distance between X
and $\tilde{X}$ is a quadratic form in X. Within the minimum distance approach the mean
squared distance between X and $\tilde{X}$ is minimized by choosing the subspace S in
an optimal way. Because of the Euclidean metric, the expected squared distance
is a function of the second-order central moments (the population covariance
matrix, which is denoted by Σ) of X. Hence it follows that the minimizing
matrix B is a function of Σ as well, which is indicated by B(Σ). However, it
should be noted that the optimal value of B is not necessarily an explicit
function of the covariance matrix Σ. For instance the characteristic vectors
and roots of Σ are implicit functions of Σ that are of importance in the
minimum distance problem.

Let us again stress the divergence of this approach from the familiar
statistical linear model as introduced in section 1. Here the linear
dependency between the components of X is not postulated beforehand. All we
try to do, is to find a random vector $\tilde{X}$ for which (2.2) holds and that
resembles the unknown X in some sense. Even when there is no indication of
some interdependence between the components of X, these operations may be
carried out as the restrictive relation holds for the vector $\tilde{X}$ to be derived
and not for X itself. All we have done so far is to indicate a method in order

to obtain the matrix $B(\Sigma)$ as a function of the parameters (e.g. the covariance matrix) of the distribution of X. No estimator of B has yet been obtained. This is the reason why this part of linear relation analysis is called the "population" part of the problem. As the optimal solution of B is obtained by minimizing an expected squared distance, these types of parameters are called minimum distance parameters.

Once the matrix B has been obtained as a function of the covariance matrix of X, we come to the "sample" part of the problem. Therefore it is assumed that a random sample $\{X_n: n=1,\ldots,N\}$ of N i.i.d. observations is at our disposal. Within the traditional linear model analysis, estimators are usually obtained by maximum likelihood techniques. This means that the observations are assumed to be generated by a specified underlying distribution, which is known up to one or more of its parameters. But this is one of the hypotheses, that have been dropped from the model specification. An alternative estimation technique is therefore applied, which is known as the method of moments. That is, if Σ is estimated by $\hat{\Sigma}$, then the estimator $\hat{B}$ of B is given by $B(\hat{\Sigma})$. Hence the argument of function $B(\Sigma)$ is replaced by its sample counterpart. Statistical properties of the estimators are obtained by using asymptotic distribution theory, which means that the sample has to consist of a large number of observations. The only assumption we need in order to obtain an asymptotically normally distributed estimator of $B(\Sigma)$, is that the dataset yields a consistent and asymptotically normal estimator of this covariance matrix. In chapter II it will be shown that when the observations are independent, and when they all have the same distribution as X, the existence of the moments of X up to the fourth-order is sufficient to find this type of estimator.

If there are problems in estimating Σ consistently, we have a so-called sample problem. It does not change the functional specification $B(\Sigma)$, as this specification is completely determined by the minimum distance solution of the population problem. Hence the problem has been partitioned into two "disjoint" components, that have been called the population part and the sample part of the problem. It is for this reason, that this method is called the Population-Sample Decomposition (PSD) approach.

The major part of this study will deal with the development of the population element, i.e. the specification of model parameters as functions of Σ. In chapter III we shall proceed from the relation as indicated in (2.2). The most general formulation is worked out in section III.1. Here the basic formulae

are derived that were already indicated in an unpublished paper by Van Praag
(1982) that laid the foundation of this PSD method (see also Van Praag, De
Leeuw, Kloek (1986)).

Note that by using the equation $B'\tilde{X} = 0$ the linear subspace S is described by
the properties (in this case the linear relations), that have to be fulfilled
by vector $\tilde{X}$. The columns of B are geometrically interpreted as the vectors
perpendicular to S. An alternative description of a subspace is given by the
so-called parametric form of the equation. Then the vector $\tilde{X}$ is defined as a
linear combination of vectors that span the subspace. Formally, a (k-p)-
dimensional linear subspace to which $\tilde{X}$ is restricted may be described by

(2.3) $\tilde{X} = AU$

with A a (k×(k-p)) matrix, the columns of which span the subspace S, and U the
(k-p)-vector of the coordinates of $\tilde{X}$ with respect to the basis formed by the
columns of A. The components of vector U are known in Factor Analysis as
"factors". The matrix A consists of the coefficients of these factors and for
this reason the elements of A are called "factor loadings". An introduction to
Factor Analysis can be found in Morrison (1978, chapter 9) and a brief review
is given in Dhrymes (1980, chapter 2)). Note that both expressions (2.2) and
(2.3) define the same linear subspace that contains $\tilde{X}$, which is the projection
of X. A similar approach to several multivariate analysis methods is given by
Keller and Wansbeek (1983) in order to obtain Maximum Likelihood estimators of
parameters in an errors-in-variables model. Relations like (2.3) are analysed
in chapter IV and formal descriptions of correspondences and differences
between the matrices B and A will be given in section IV.2. By assuming
specific forms for the Euclidean metric with respect to which the distances
are minimized and by placing restrictions on the parameter matrices, solutions
will be obtained, the sample counterparts of which correspond to some well-
known multivariate statistics. These correspondences are indicated by the
titles of the sections of chapter III and IV. However, one must bear in mind
that those names are merely chosen to symbolize the equivalence between the
form of the minimum distance parameters and the multivariate statistics with
these names. The parameters are nòt generated by adopting the linear model for
true as in "classical" multivariate analysis. In our case the model is
restricted to the assumption of a random sample with observations, the moments
of which are finite up to the fourth-order. The distance minimizing parameters
are obtained as functions of the covariance matrix of X and estimators are

obtained by replacing the covariance matrix of X by its sample analogue to serve as an argument of the distance minimizing parameter matrix. Formulas (2.2) and (2.3) give relations, that hold exactly for the vector $\tilde{X}$. In general these relations do not hold for X itself. The question arises: to what extent do we value the parameter matrices A and B when the relations found are not the real model? As the expected squared distance between X and $\tilde{X}$ is minimized, it is obvious that (2.2) and (2.3) may be referred to as the "best fitting" linear subspaces for the distribution of X. However, how well do they fit? An answer to this question is given in chapter V, where the deviation of $\tilde{X}$ from X (that is the expected squared distance between $\tilde{X}$ and X) relative to the expected squared length of X is defined as a goodness-of-fit measure. The goodness-of-fit measure is used to evaluate how well the distribution of X can be forced into a linear subspace of lower dimension. These measures of linear association between the components of X may be compared with the multiple correlation coefficient, that measures the degree-of-explanation in the linear regression model, or with likelihood ratios and information criteria that are reviewed by Akaike (1981) and Atkinson (1981).

Chapters III, IV and V consist mainly of lengthy matrix operations. Therefore the results of these chapters are briefly summarized in chapter VI, where the relations between the various sections are given in a schematic overview. From chapter II, that deals with the estimation of the population parameters and stands for the sample part of the PSD approach, it will become clear that the PSD method has some attractive computational characteristics. However, as the asymptotic distribution of the estimators requires the computation of all fourth-order central moments of X, something has to be said about the requirement of computer space in order to store this large amount of statistics. This will be done in chapter VII. Also a general description of how to organize a computer program (in PASCAL) of the PSD method will be given here. This study ends with an appendix on the main results of matrix algebra and matrix differentiation theory, that is essential for the further pursuit of our objectives. These results will be referred to when they are used in the derivations by means of the formula number, starting with an A, given between brackets.

II. The Estmation of Linear Relations; The Sample Part of PSD

II.1 Method of moments and asymptotic distribution theory

This chapter is devoted to the estimation of parameters, that are differentiable functions of the moments of the random vector X. Before we come to the estimation technique (in section 2), first some notational conventions and preliminary statistical theory will be given. For a more thorough and general explanation on (asymptotic) estimation theory one is referred to statistical textbooks like Mood, Graybill and Boes et al. (1974), Serfling (1980) and White (1984).

Assume that the first- and second-order moments of the vector X exist (that is, they are finite). They are denoted by the population mean

$$(1.1) \qquad \mu = E(X) \qquad\qquad (k \times 1)$$

and the population covariance matrix

$$(1.2) \qquad \Sigma = E(X-\mu)(X-\mu)' \qquad\qquad (k \times k).$$

It is also assumed that a set of N observations on X is observed, yielding a sample $\{X_n: n=1,\ldots,N\}$ of independent and identically distributed (i.i.d.) random vectors. Consistent and asymptotically unbiased estimators of μ and Σ are given by their sample counterparts

$$(1.3) \qquad \hat{\mu} = \frac{1}{N} \sum_{n=1}^{N} X_n \qquad\qquad (k \times 1)$$

$$(1.4) \qquad \hat{\Sigma} = \frac{1}{N} \sum_{n=1}^{N} (X_n-\hat{\mu})(X_n-\hat{\mu}) \qquad\qquad (k \times k).$$

The estimators are functions of the random observations and can therefore be seen as a vector or matrix of random variables themselves. Efficiency statements about estimators are made in terms of variances and standard deviations of the estimators. Therefore, knowledge about the distribution of the estimators is important. Let us consider the distribution of $\hat{\mu}$ more closely. Naturally it is not possible to present any definite statements on the distribution of the sample mean, as the distributional properties of the observations are unknown. However, if it is assumed that the dataset consists of many observations (or in other words: N is assumed to be large), then the distribution of $\hat{\mu}$ can frequently be approximated by a normal distribution according to the central limit theorem. Here the central limit theorem is presented in the most widely known version.

Theorem (the Lindeberg–Lévy central limit theorem; CLT)
Let $\{X_n : n=1,\ldots,N\}$ be a set of N i.i.d. random vectors with mean μ and covariance matrix Σ and let $\hat{\mu}$ be defined as in (1.3). Then

$$(1.5) \qquad \sqrt{N}(\hat{\mu} - \mu) \xrightarrow{D} \mathbf{N}(0, \Sigma).$$

This implies that the estimator $\hat{\mu}$ of the population mean μ is asymptotically normally distributed with covariance matrix $\frac{1}{N}\Sigma$

$$(1.6) \qquad \hat{\mu} \sim \mathbf{AN}(\mu, \tfrac{1}{N}\Sigma).$$

where $\mathbf{AN}$ stands for asymptotic normality. Proofs of the univariate case of CLT (k=1) are provided in many textbooks and are based on the convergence of the characteristic function of $\sqrt{N}(\hat{\mu} - \mu)$ to the characteristic function of the normal distribution, if $N \to \infty$. The multivariate extension of CLT may be derived by using the Cramér–Wold device, that allows the problem of convergence of multivariate distribution functions to be reduced to that of convergence of univariate distribution functions (see for instance Rao (1973, p. 128) and Serfling (1980, p. 18)).
In fact, the CLT is concerned with the convergence in distribution of sums of random vectors under the assumption that these vectors are i.i.d. with finite covariance matrix. Generalizations of the Lindeberg–Lévy version can be obtained for not necessarily identically distributed observations or a random number of summands (hence N is stochastic). For alternative versions of the central limit theorems and their proofs, see for instance Feller (1966, chapters XV and XVI), Rao (1973, section 20.5) or Serfling (1980, section 1.9)

Many times, the population parameter vector to be estimated is a known function of the moments of X. For instance let θ be a parameter ($\ell \times 1$) vector that can be written as a function of μ. Denote this by defining a function on (a subset of) $\mathbb{R}^k$ with function values in $\mathbb{R}^\ell$ as follows

(1.7) $\theta = \phi(\mu)$.

The estimator $\hat{\theta}$, that is obtained by replacing in function ϕ the population moments by the corresponding sample moments, is called the "method of moments estimator" (MME). Analogous to (1.7) we may write

(1.8) $\hat{\theta} = \phi(\hat{\mu})$.

The method of moments, introduced by Pearson (1894), has enjoyed wide appeal because of its simplicity. Further, given the convergence in distribution of $\sqrt{N}(\hat{\mu} - \mu)$ to the normal distribution and the continuity of ϕ with probability 1, the MME is consistent and asymptotically normal. The parameters of the limiting distribution of $\hat{\theta}$ can be derived by application of a theorem that is known in the literature under the name "delta method". Here we shall use the delta method in the following form.

Theorem (delta method)
Let $\hat{\mu}$ be a random (k×1) vector, asymptotically normally distributed as denoted in (1.5). Let ϕ be a function defined on an open neighbourhood of μ with values in R^ℓ and differentiable in μ. When the partial derivatives of ϕ with respect to its arguments, evaluated in μ are given in the (ℓ×k) gradient matrix $\nabla(\phi;\mu)$ of ϕ in μ and do not all equal zero, then

(1.9) $\sqrt{N}(\phi(\hat{\mu}) - \phi(\mu)) \xrightarrow{D} N(0, [\nabla(\phi;\mu)]\Sigma[\nabla(\phi;\mu)]')$.

The essence of this asymptotic method is approximation. As careful use of approximations requires attention to their accuracy or order, first the "little-o" notation will be defined.

Definition (little-o)
Let a_N be a sequence of scalars and let m_N denote a sequence of vectors (N = 1,2,...). Then $m_N = o(a_N)$ (read: m_N is little o of a_N) means that

(1.10) $\lim_{N\to\infty} \frac{m_N}{a_N} = 0$.

When the vector sequence m_N has a stochastic character, the little-o notation is generalized in probability terms as follows

Definition (stochastic little-o_p)

$m_N = o_p(a_N)$ (or, equivalently, $\dfrac{m_N}{a_N} = o_p(1)$) if for every $\varepsilon > 0$

(1.11) $\displaystyle\lim_{N\to\infty} P\left(\left| \frac{m_N}{a_N} \right| \leq \varepsilon \right) = 1,$

or $\dfrac{m_N}{a_N} \overset{P}{\to} 0$ (convergence in probability). In econometric literature this is frequently denoted as $\displaystyle\plim_{N\to\infty} \frac{m_N}{a_N} = 0.$

The generalization of the o-notation for non-stochastic sequences to the o_p-notation for stochastic sequences and its applications in statistical approximation theory is based on work of Chernoff (1956) and is extended by Pratt (1959).

Proof (of the delta method)

Let m_N by a sequence of non-random ($k \times 1$) vectors and $\displaystyle\lim_{N\to\infty} m_N = \mu$. In what follows m_N serves as an argument for the function ϕ. Consider the following first-order Taylor expansion of ϕ about μ

(1.12) $\phi(m_N) = \phi(\mu) - [\nabla(\phi;\mu)](m_N - \mu) + o(\|m_N - \mu\|)$ for $N \to \infty$.

Replacing in (1.12) m_N by the consistent estimator $\hat{\mu}$ of μ, we obtain

(1.13) $\phi(\hat{\mu}) - \phi(\mu) = [\nabla(\phi;\mu)](\hat{\mu} - \mu) + o_p(1/\sqrt{N})$ for $N \to \infty$.

Multiplication of (1.13) by $\sqrt{N}$ yields

(1.14) $\sqrt{N}(\phi(\hat{\mu}) - \phi(\mu)) = [\nabla(\phi;\mu)]\sqrt{N}(\hat{\mu} - \mu) + o_p(1)$ for $N \to \infty$.

From the CLT it follows that the limiting distribution of the first term of the right hand side of (1.14) is given by

$$(1.15) \qquad [\nabla(\phi;\mu)]\sqrt{N}(\hat{\mu} - \mu) \overset{D}{\to} \mathbb{N}(0, \ [\nabla(\phi;\mu)]\Sigma[\nabla(\phi;\mu)]').$$

As $o_p(1)$ stands for a term that converges in probability to zero, (1.14) implies that $\sqrt{N}(\phi(\hat{\mu}) - \phi(\mu))$ and $[\nabla(\phi;\mu)]\sqrt{N}(\hat{\mu} - \mu)$ have the same limiting distribution. Combination of this fact and (1.15) yields (1.9), which completes the proof. ◊

For more extensive information about the approximation of (random) functions by means of the attractive o- and o_p-notation, the reader is referred to Bishop, Fienberg and Holland (1975, chapter 14). An alternative proof of the delta method is given by Serfling (1980, p. 118). Univariate versions of the delta method can for instance be found in Rao (1973, p. 388) and Bishop, Fienberg and Holland, (1975, p. 486).

Application of the delta method to the MME given in (1.8) yields an approximation of its distribution

$$(1.16) \qquad \sqrt{N}(\hat{\theta} - \theta) \overset{D}{\to} \mathbb{N}(0, \ [\nabla(\phi;\mu)]\Sigma[\nabla(\phi;\mu)]'),$$

where it is assumed that ϕ is differentiable in μ and its partial derivatives do not all equal zero. Consequently a consistent estimator of the covariance matrix of $\hat{\theta}$ is given by

$$(1.17) \qquad \frac{1}{N} \ [\nabla(\phi;\mu)]_{\hat{\mu}=\mu} \ \hat{\Sigma} \ [\nabla(\phi;\mu)]'_{\hat{\mu}=\mu} \qquad\qquad (\ell\times\ell).$$

A typical criticism of the method of moments is that MME estimators are generally not asymptotically efficient. Therefore modified method of moment approaches, that improve asymptotic efficiency, have been introduced. Soong (1969) proposed to use optimal linear combinations of the sample moments as alternative moments estimators and the large sample properties of a class of so-called generalized MME are discussed by Hansen (1982). However, it should be noticed that when having no prior information on the distribution of the sample observations, it often is the only estimation technique that can be applied.

II.2 Asymptotic estimation of covariance functions

The population parameters, introduced in the chapters to come, typically are
functions of the population covariance matrix Σ. Therefore these parameter
vectors are called "covariance functions". We are interested in the large
sample properties of the MME of these covariance functions, which can be
obtained by means of the delta method. Distribution-free estimators for
structural models have also been discussed in the context of analysis of
covariance structures (Browne (1974, 1984), Bentler (1983)) and applications
of this technique to psychometric and econometric models have been considered
by Bentler and Dijkstra (1983). Applications of the delta method, in order to
derive the limiting distribution of the estimated covariance function, should
be preceded by verification of the asymptotic normality of the sample
covariance matrix $\hat{\Sigma}$. As $\hat{\Sigma}$ is defined to be the sample mean of $(X_n-$
$\mu)(X_n-\mu)'$, the covariance matrix of this random matrix is needed. Let us first
consider the vectorized version of its population analogue: $vec((X-\mu)(X-\mu)')$.
This is a $(k^2 \times 1)$ vector, the covariance matrix of which is defined by

$$(2.1) \quad V = E[(vec((X-\mu)(X-\mu)'))(vec((X-\mu)(X-\mu)'))'] - (vec\Sigma)(vec\Sigma)' \quad (k^2 \times k^2).$$

In what follows it is assumed that this matrix is finite, which means that the
moments of X are assumed to exist up to the fourth order. Now a theorem about
the limiting distribution of $\hat{\Sigma}$ can be formulated as follows

Theorem (asymptotic normality of $\hat{\Sigma}$)
Let $\{X_n : n=1,\ldots,N\}$ be a set of i.i.d. random vectors with mean μ, covariance
matrix Σ and V as defined in (2.1). Let $\hat{\Sigma}$ be the sample covariance matrix
defined in (1.4), then

$$(2.2) \quad \sqrt{N}(vec\hat{\Sigma} - vec\Sigma) \xrightarrow{D} N(0, V).$$

Proof
Let the sample mean of the matrix of cross-products of $(X_n-\mu)$ be defined by

$$(2.3) \quad \tilde{\Sigma} = \frac{1}{N} \sum_{n=1}^{N} (X_n-\mu)(X_n-\mu)' \quad (k \times k)$$

where the sample mean $\hat{\mu}$ in $\hat{\Sigma}$ has been replaced by population mean μ. The random vector of interest may be decomposed into two terms as

$$(2.4) \qquad \sqrt{N}(vec\hat{\Sigma} - vec\Sigma) = \sqrt{N}(vec\hat{\Sigma} - vec\tilde{\Sigma}) + \sqrt{N}(vec\tilde{\Sigma} - vec\Sigma).$$

The two terms on the right-hand side of (2.4) will be considered separately. As $vec\tilde{\Sigma}$ is the sample mean of i.i.d. random vectors $vec((X_n-\mu)(X_n-\mu)')$ with mean $vec\Sigma$ and covariance matrix V, the limiting distribution of the second term of the right hand side of (2.4) is given by (application of CLT)

$$(2.5) \qquad \sqrt{N}(vec\tilde{\Sigma} - vec\Sigma) \overset{D}{\to} N(0, V).$$

The difference between $vec\hat{\Sigma}$ and $vec\tilde{\Sigma}$ is given by

$$(2.6) \qquad vec\hat{\Sigma} - vec\tilde{\Sigma} = -vec((\hat{\mu}-\mu)(\hat{\mu}-\mu)') = -(\hat{\mu}-\mu) \otimes (\hat{\mu}-\mu) \qquad (k^2 \times 1).$$

As $\sqrt{N}(\hat{\mu} - \mu)$ converges in distribution (asymptotically normal $N(0, \Sigma)$) and $(\hat{\mu} - \mu)$ converges in probability to zero ($\hat{\mu}$ is a consistent estimator of μ), (2.6) yields

$$(2.7) \qquad \sqrt{N}(vec\hat{\Sigma} - vec\tilde{\Sigma}) \overset{P}{\to} 0.$$

Combination of (2.4), (2.5) and (2.7) yields asymptotic equivalence between $\hat{\Sigma}$ and $\tilde{\Sigma}$, which completes the proof. (Some properties concerning the combination of convergence in probability and convergence in distribution can be found in Rao (1974, p. 122)). ◇

The main problem in the preceding proof is that the estimator $\hat{\Sigma}$ is not a sum of vectors $vec((X_n-\mu)(X_n-\mu)')$, but of $vec((X_n-\hat{\mu})(X_n-\hat{\mu})')$ (for $n=1,\ldots,N$). Hence, the parameter vector μ is replaced by its estimator $\hat{\mu}$. It appears that this does not influence the limiting distribution of the sample covariance matrix. More general results on the asymptotic effect of substituting estimators for parameters in statistics can be found in Randles (1982) and Pierce (1982). In fact, this theorem on asymptotic normality of $\hat{\Sigma}$ is a special case of the class of estimators they found not to be affected by this kind of substitutions.

If the $(\ell \times 1)$ parameter vector θ can be written as a function ϕ defined on (a subset of) R^{k^2} with function values in R^ℓ, which is differentiable in Σ, then the delta method can be applied and the asymptotic distribution of the method of moments estimator $\hat{\theta} = \phi(\hat{\Sigma})$ is obtained, yielding

$$(2.8) \qquad \sqrt{N}(\hat{\theta} - \theta) \overset{D}{\to} N(0, \ [\nabla(\phi;\Sigma)]V[\nabla(\phi;\Sigma)]')$$

with $\nabla(\phi;\Sigma)$ the $(\ell \times k^2)$ gradient matrix $\dfrac{\partial\phi}{\partial(vec\Sigma)'}$, consisting of the partial derivatives of function ϕ with respect to the vectorized matrix of its arguments, evaluated in Σ. A consistent estimator of the covariance matrix of the limiting distribution is obtained by replacing Σ and V by their sample analogues. Hence the approximated covariance of $\hat{\theta}$ is consistently estimated by

$$(2.9) \qquad \frac{1}{N} [\nabla(\phi;\Sigma)]_{\Sigma=\hat{\Sigma}} \ \hat{V} \ [\nabla(\phi;\Sigma)]_{\Sigma=\hat{\Sigma}} \qquad\qquad (\ell \times \ell)$$

with

$$(2.10) \qquad \hat{V} = \frac{1}{N} \sum_{n=1}^{N} (vec((x_n-\hat{\mu})(x_n-\hat{\mu})'))(vec((x_n-\hat{\mu})(x_n-\hat{\mu})'))' - (vec\hat{\Sigma})(vec\hat{\Sigma})'$$
$$(k^2 \times k^2).$$

One remark about the preceding matrix manipulations should be made. Due to symmetry of $\hat{\Sigma}$, $\frac{1}{2}k(k-1)$ of the k^2 elements of $vec\hat{\Sigma}$ are redundant. Consequently, the covariance matrix V of the limiting distribution as denoted in (2.2) is singular, but this does not imply singularity of (2.9). This could be avoided by eliminating the infra-diagonal elements of $\hat{\Sigma}$. That is, instead of stacking all the elements of $\hat{\Sigma}$, only the different elements of $\hat{\Sigma}$ are vectorized, yielding a vector of length $\frac{1}{2}k(k+1)$. However, ignoring the symmetry of $\hat{\Sigma}$ in both its limiting distribution and in the gradient matrix $\nabla(\phi;\Sigma)$ makes no difference for the limiting distribution of $\hat{\theta}$ as denoted in (2.8). As the vectorization of the upper triangular part of symmetric matrices would require additional notations, while the derivations of the required limiting distributions essentially remain the same, the use of these more economic expressions shall here be omitted. (References to related studies can be found in the appendix).

From the preceding formulas it may be clear, that an essential role is played by the matrix V, which denotes the covariance matrix of $\text{vec}((X-\mu)(X-\mu)')$. Therefore it seems wise to provide more insight into the construction of this large matrix. From definition (2.1) it follows that every component of V is composed of two terms, viz. a fourth-order central moment, that is a component of the matrix $E[(\text{vec}((X-\mu)(X-\mu)'))(\text{vec}((X-\mu)(X-\mu)'))']$, and the product of two (co)variances, that is an element of the matrix $(\text{vec}\Sigma)(\text{vec}\Sigma)'$. It would be straightforward to place the fourth-order central moments in a 4-dimensional (k×k×k×k) matrix, just as the second-order moments are ordered in the 2-dimensional (k×k) matrix Σ. However, as V stands for the covariances of a vectorized form of $(X-\mu)(X-\mu)'$ the elements of the (k×k×k×k) fourth-order moments matrix are rearranged into a 2-dimensional $(k^2 \times k^2)$ matrix. More precise statements about the grouping of the fourth-order moments can be made after defining a notation for the components of vector X. Let the components of X, mean vector μ and covariance matrix Σ be defined by

(2.11) $\quad X = (X_1,\ldots,X_k)'$

$\mu = (\mu_1,\ldots,\mu_k)' \qquad$ with $\mu_i = E(X_i) \qquad\qquad i=1,\ldots,k$

$\Sigma = [\sigma_{ij}] \qquad\qquad$ with $\sigma_{ij} = E(X_i-\mu_i)(X_j-\mu_j) \quad i,j=1,\ldots,k.$

The $(k^2 \times k^2)$ matrix V can be considered as a (k×k) matrix of (k×k) blocks. The $(h,\ell)^{th}$ block is defined by

(2.12) $\quad V_{.h,.\ell} = E((X-\mu)(X_h-\mu_h))((X-\mu)(X_\ell-\mu_\ell))' - \Sigma_{.h}\Sigma'_{.\ell} \qquad\qquad$ (k×k)

with $\Sigma_{.h}$ the hth column of matrix Σ.

Now the $(i,j)^{th}$ element of the $(h,\ell)^{th}$ block of V is given by

(2.13) $\quad V_{ih,j\ell} = E(X_i-\mu_i)(X_h-\mu_h)(X_j-\mu_j)(X_\ell-\mu_\ell) - \sigma_{ih}\sigma_{j\ell}.$

In other words $V_{ih,j\ell}$ stands for the $((h-1)k+i,(\ell-1)k+j)^{th}$ element of V. Introducing a notation for the fourth-order central moments by

(2.14) $\quad \upsilon_{ihj\ell} = E(X_i-\mu_i)(X_h-\mu_h)(X_j-\mu_j)(X_\ell-\mu_\ell),$

formula (2.13) may be abbreviated as

(2.15) $\quad V_{ih,j\ell} = \upsilon_{ihj\ell} - \sigma_{ih}\sigma_{j\ell}.$

Sometimes statistical analyses are carried out after the observations have
been standardized. That is, X is normalized such that the variances of the k
components of X all equal 1. This is done in order to make all population
parameters independent of the units in which the various variables are
measured. The covariance matrix of a standardized random vector is called the
correlation matrix, say R. The diagonal elements of R equal unity and the off-
diagonal elements stand for the correlation coefficients. In terms of
covariance matrix components the $(i,j)^{th}$ element of R is given by

$$(2.16) \qquad \rho_{ij} = \frac{\sigma_{ij}}{\sqrt{\sigma_{ii}}\sqrt{\sigma_{jj}}}.$$

Now the estimators to be considered are not defined as functions of the sample
covariance matrix $\hat{\Sigma}$, but as functions of the sample correlation matrix $\hat{R}$, the
elements of which are analogous to (2.16) given by

$$(2.17) \qquad \hat{\rho}_{ij} = \frac{\hat{\sigma}_{ij}}{\sqrt{\hat{\sigma}_{ii}}\sqrt{\hat{\sigma}_{jj}}}.$$

Consequently for application of the delta method in case of standardized
observations, the asymptotic distribution of the vectorized sample correlation
matrix vec$\hat{R}$ is needed. Note that the correlation matrix R itself can be
considered as a covariance function $R(\Sigma)$. Hence vec$\hat{R}$ is asymptotically
normally distributed and its limiting distribution can be obtained by
application of the delta method with function R as defined in (2.16). The
resulting limiting distribution will be given in a theorem, that uses the
following notational conventions. The Kronecker delta δ_{ij} is a 0,1 variable
defined as

$$(2.18) \qquad \delta_{ij} = \begin{cases} 1 & \text{if } i = j \\ 0 & \text{if } i \neq j \end{cases}$$

and a generalization of the correlation coefficients from second-order moments to fourth-order moments is given by

$$(2.19) \qquad \rho_{ihj\ell} = \frac{\upsilon_{ihj\ell}}{\sqrt{\sigma_{ii}}\sqrt{\sigma_{hh}}\sqrt{\sigma_{jj}}\sqrt{\sigma_{\ell\ell}}}.$$

Theorem (asymptotic normality of vec$\hat{R}$)

Let $\{X_n: n=1,\ldots,N\}$ be a set of N i.i.d. random vectors with mean μ, covariance matrix Σ, correlation matrix R and V as defined in (2.1). Let $\hat{R}$ be the sample correlation matrix defined in (2.17), then

$$(2.20) \qquad \sqrt{N}(\text{vec}\hat{R} - \text{vec}R) \xrightarrow{D} N(0, W)$$

where the $((h-1)k+i,(\ell-1)k+j)^{th}$ element of W equals

$$(2.21) \qquad (1-\delta_{ih})(1-\delta_{j\ell})(\rho_{ihj\ell}-\tfrac{1}{2}\rho_{ih}(\rho_{iij\ell}+\rho_{hhj\ell}) - \tfrac{1}{2}\rho_{j\ell}(\rho_{ihjj}+\rho_{ih\ell\ell})$$

$$+ \tfrac{1}{4}(\rho_{ih}\rho_{j\ell}(\rho_{iijj}+\rho_{hhjj}+\rho_{ii\ell\ell}+\rho_{hh\ell\ell})).$$

Note that because of factor $(1-\delta_{ih})(1-\delta_{j\ell})$ the (co)variances on positions in W where $i=h$ or $j=\ell$ equal zero. This is logical, as these elements of W stand for the covariances between two sample correlations of which at least one of them equals the diagonal element 1 of $\hat{R}$. Furthermore it should be noted that this result is, in contrast to other formulae, not given in matrix notation, as the introduction of necessary additional symbols is not worth-while.

Proof

According to the first theorem of this section, vec$\hat{\Sigma}$ is asymptotically normal distributed (see (2.2)) with the elements of V given in (2.13). Application of the delta method with R as a function of Σ yields

$$(2.22) \qquad \sqrt{N}(\text{vec}\hat{R} - \text{vec}R) \xrightarrow{D} N(0, [\nabla(R;\Sigma)]V[\nabla(R;\Sigma)]')$$

with $\nabla(R;\Sigma)$ the gradient matrix

(2.23) $\nabla(R;\Sigma) = \dfrac{\partial(vecR)}{\partial(vec\Sigma)'}$ $(k^2 \times k^2)$.

Note that in fact the gradient matrix should be denoted by $\nabla(vecR;vec\Sigma)$, however for notational convenience, notation (2.23) will be used. More about the notational conventions concerning the derivatives of matrices can be found in the last part of the appendix.

The asymptotic covariance matrix obtained after pre- and post-multiplication of V by $\nabla(R;\Sigma)$ and its transpose will be called W. The $((h-1)k+i,(\ell-1)k+j)^{th}$ element of $\nabla(R;\Sigma)$ is indicated by $[\nabla(R;\Sigma)]_{ih,j\ell}$ and equals the partial derivative

(2.24) $\dfrac{\partial \rho_{ih}}{\partial \sigma_{j\ell}} = \begin{cases} \dfrac{\rho_{ih}}{\sigma_{ih}} & \text{if } i \neq h,\ j=i,\ \ell=h \\[2mm] -\dfrac{\rho_{ih}}{2\sigma_{ii}} & \text{if } i \neq h,\ j=i,\ \ell=i \\[2mm] -\dfrac{\rho_{ih}}{2\sigma_{hh}} & \text{if } i \neq h,\ j=h,\ \ell=h \\[2mm] 0 & \text{if otherwise.} \end{cases}$

Using the Kronecker delta, (2.24) can be rewritten as

(2.25) $[\nabla(R;\Sigma)]_{ih,j\ell} = (1-\delta_{ih})\rho_{ih}\left(\dfrac{\delta_{ij}\delta_{h\ell}}{\sigma_{ih}} - \dfrac{\delta_{ij}\delta_{i\ell}}{2\sigma_{ii}} - \dfrac{\delta_{hj}\delta_{hi}}{2\sigma_{hh}}\right).$

Consequently, the elements of V as well as the elements of $\nabla(R;\Sigma)$ are known and all we need to do in order to obtain the elements of W is to perform the pre- and post-multiplication of V by $\nabla(R;\Sigma)$ and its transpose. Element $((h-1)k+i,(\ell-1)k+j)$ of the matrix $[\nabla(R;\Sigma)]V$ obtained after pre-multiplication equals

(2.26) $\displaystyle\sum_{r=1}^{k}\sum_{s=1}^{k}[\nabla(R;\Sigma)]_{ih,rs}\, V_{rs,j\ell}$

$$= \sum_{r=1}^{k} \sum_{s=1}^{k} (1-\delta_{ih})\rho_{ih}(\frac{\delta_{ir}\delta_{hs}}{\sigma_{ih}} - \frac{\delta_{ir}\delta_{is}}{2\sigma_{ii}} - \frac{\delta_{hr}\delta_{hs}}{2\sigma_{hh}})(V_{rs,j\ell} - \sigma_{rs}\sigma_{j\ell})$$

$$= (1-\delta_{ih})\rho_{ih}(\frac{\upsilon_{ihj\ell}}{\sigma_{ih}} - \frac{\upsilon_{iij\ell}}{2\sigma_{ii}} - \frac{\upsilon_{hhj\ell}}{2\sigma_{hh}}).$$

Post-multiplication of $[\nabla(R;\Sigma)]V$ by $[\nabla(R;\Sigma)]'$ yields the matrix W; its $((h-1)k+i,(\ell-1)k+j)^{th}$ element equals

$$(2.27) \quad \sum_{r=1}^{k} \sum_{s=1}^{k} (1-\delta_{ih})\rho_{ih}(\frac{\upsilon_{ihrs}}{\sigma_{ih}} - \frac{\upsilon_{iirs}}{2\sigma_{ii}} - \frac{\upsilon_{hhrs}}{2\sigma_{hh}})\rho_{j\ell}(\frac{\delta_{j\ell}\delta_{\ell s}}{\sigma_{j\ell}} - \frac{\delta_{jr}\delta_{js}}{2\sigma_{jj}} - \frac{\delta_{\ell r}\delta_{\ell s}}{2\sigma_{\ell\ell}})$$

$$= (1-\delta_{ih})(1-\delta_{j\ell})\rho_{ih}\rho_{j\ell}((\frac{\upsilon_{ihj\ell}}{\sigma_{ih}\sigma_{j\ell}} - \frac{\upsilon_{iij\ell}}{2\sigma_{ii}\sigma_{j\ell}} - \frac{\upsilon_{hhj\ell}}{2\sigma_{hh}})$$

$$- (\frac{\upsilon_{ihjj}}{2\sigma_{ih}\sigma_{jj}} - \frac{\upsilon_{iijj}}{4\sigma_{ii}\sigma_{jj}} - \frac{\upsilon_{hhjj}}{4\sigma_{hh}\sigma_{jj}})$$

$$- (\frac{\upsilon_{ih\ell\ell}}{2\sigma_{ih}\sigma_{\ell\ell}} - \frac{\upsilon_{ii\ell\ell}}{4\sigma_{ii}\sigma_{\ell\ell}} - \frac{\upsilon_{hh\ell\ell}}{4\sigma_{hh}\sigma_{\ell\ell}})).$$

Substitution of (2.19) in (2.27) yields expression (2.21). ◊

These results were already shown by Shu (1949) and more recently Steiger and Hakstian have paid attention to the asymptotic distribution of correlation coefficients (1982). De Leeuw (1983) illustrated the use of correlation coefficients in discrete multinormal models and their application to Spearman's (1904) two-factor model.

This chapter ends by explaining the title "Population-Sample Decomposition" from the distribution formulas derived in this section and illustrating its practical attractiveness. First, we recapitulate the principal formulas that are used to estimate the population parameters of the next three chapters.

$$(2.28) \quad \sqrt{N}(\text{vec}\hat{\Sigma} - \text{vec}\Sigma) \overset{D}{\rightarrow} N(0, V)$$

(2.29) $\qquad \sqrt{N}(\hat{\theta} - \theta) \overset{D}{\to} \mathbf{N}(0, [\nabla(\phi;\Sigma)]V[\nabla(\phi;\Sigma)]')$

with $\theta = \phi(\Sigma)$ and $\hat{\theta} = \phi(\hat{\Sigma})$.

Let us first examine statement (2.28) about the limiting distribution of vec$\hat{\Sigma}$. It says that $\hat{\Sigma}$ is an asymptotically normal estimator of Σ. We have seen that this requirement is fulfilled when it is assumed that the dataset $\{X_n: n=1,\dots,N\}$ consists of i.i.d. observations with finite fourth-order moments. These were just the demands needed to apply the Lindeberg-Lévy CLT. However, there exist many versions of the central limit theorem and they ensure asymptotic normality of the sample covariance matrix under far more general conditions. The practice of deriving an asymptotic normal estimator of Σ and the corresponding covariance matrix of the limiting distribution completely depends on the assumptions about the distribution of the sample observations and is independent of the population parameter θ. It is called the "sample part" of the estimation problem. In order to apply the delta method, the parameter vector θ needs to be specified as a function $\theta = \phi(\Sigma)$ of the population covariance matrix Σ and its partial derivatives $\nabla(\phi;\Sigma)$ have to be computed as well. The problem of specifying the function ϕ and its derivatives does not depend on the sample observations and stands for the "population part" of the problem. As the estimation of Σ on one hand and the derivation of ϕ and its derivatives on the other hand are two "disjoint" problems, this method is called the Population-Sample Decomposition approach. This dichotomy manifests itself especially in the covariance matrix of the limiting distribution (see (2.29)): V depends on the nature of the sampling procedure and $\nabla(\phi;\Sigma)$ represents the parameter characteristics.

Examples of non-ideal samples (that is observations not in conformity with the demands of the Lindeberg-Lévy CLT) are for instance collections of observations with missing data, selective samples and datasets of which the observations are mutually correlated. Consistent and asymptotically normal estimators of the second-order moments with the corresponding covariance matrix are, for some of these cases, given in Van Praag, Wesselman (1984) and in Van Praag, Dijkstra, Van Velzen (1985). The first paper analyses the solution of the incomplete data problem by means of the hot-deck method, i.e. the unknown value is set equal to the observed value in the last preceding observation (see also Bailar and Bailar (1979), Sande (1979) and Ford (1980)). The latter paper extends the pairwise-deletion method (see also Haitovsky

(1968)), that is a standard option in many statistical software packages. By a
selective sample is meant a set of observations, that does not represent the
distribution of X in the population. This causes the traditional estimators of
the population moments to be biased (selectivity bias). Asymptotically
unbiased and normal estimators are given in Wesselman, Van Praag (1984).
In contrast with the less informative non-ideal sample, it can also be that,
in addition to the i.i.d. hypothesis, we have some prior knowledge on the
distributional form of the data. For instance, when the sample observations
are not only known to be i.i.d., but also normally distributed, then the
relation between the fourth-order central moments and the elements of the
matrix Σ is given by

(2.30) $\upsilon_{ihj\ell} = \sigma_{ih}\sigma_{j\ell} + \sigma_{ij}\sigma_{h\ell} + \sigma_{i\ell}\sigma_{hj}.$

Substitution of (2.30) in (2.15) yields that under the normality assumption
the $((h-1)k+i,(\ell-1)k+j)^{th}$ element of V equals

(2.31) $V_{ih,j\ell} = \sigma_{ij}\sigma_{h\ell} + \sigma_{i\ell}\sigma_{hj}.$

Rewriting (2.31) in matrix notation we obtain

(2.32) $(\Sigma \otimes \Sigma)(I_{k^2} + P_{k,k})$ $(k^2 \times k^2)$

with the Kronecker product $\otimes$ and permutation matrix $P_{k,k}$ defined in the
appendix (see (A.2) and (A.15). A larger family of distributions, that
includes the normal distribution as a special case, is the class of so-called
elliptical distributions. These distributions, of which the equi-probability
contours have the same elliptical shape as the normal, provide attractive
extensions to multivariate normality. Their fourth-order moments are simple
functions of the covariances and the kurtosis of the distribution. (The
kurtosis equals zero in case of normality). Detailed studies of elliptical
distributions are given by Kelker (1970) and Muirhead (1980) and some
applications to the PSD method may be found in Wesselman, Van Praag (1987) and
Van Praag, Wesselman (1986).
The development of the thoughts on the separation of population problems and
sample problems was initialized by Van Praag (1982). Van Praag, De Leeuw,
Kloek (1986) extended these ideas and related the PSD approach to existing

estimation methods of linear models. The population problem is generalized by Van Praag and Koster (1984) by posing additional constraints on the population parameters and a Lagrangian test of the validity of the constraints is discussed by them.

In the next three chapters of this study the theory of deriving the population parameters as minimum distance parameters will be elaborated and the results will be related to some well-known multivariate statistical methods. In order to apply the method of moments and the delta method, they will be obtained as functions of the covariance matrix and every parameter vector will be accompanied by the matrix of partial derivatives.

III. Principal Relations

III.1 Basic formulation of principal relations

In the analysis of principal relations (PR) we shall proceed from the
existence of a random (k×1) vector X, with an unknown distribution function,
the moments of which are assumed to exist up to the fourth-order. In order to
keep the notation convenient (and without loss of generality) the first-order
moments (the expectations) of X are assumed to be zero. Consequently the
covariance matrix of X is given by

$$(1.1) \qquad \Sigma = E(XX') \qquad\qquad (k \times k),$$

This matrix is assumed to be finite and non-singular. Furthermore the fourth-
order moments are assumed to exist, in order to make statistical inferences
about the estimator $\hat{\Sigma}$ possible.

The basic idea of PR is to find a linear subspace in R^k that fits the
distribution of X "as close as possible". Two geometrical aspects should be
considered in advance. Firstly the dimension of the linear subspace and
secondly the metric that defines the words "as close as possible" in
mathematical terms. Let ξ be a vector in R^k. The dimension of the linear
subspace to which ξ belongs is defined by the number of independent linear
relations. That is, when ξ is assumed to fulfill p, say, independent linear
relations

$$(1.2) \qquad B'\xi = 0,$$

with B a (k×p) matrix of rank p, then ξ is situated in the (k-p)-dimensional
linear subspace orthogonal (with respect to the simple Euclidean metric) to
the subspace spanned by the columns of B. In what follows the subspace spanned
by the columns of a matrix (here matrix B) will be denoted by span(B). Its
orthogonal complement will be denoted by O(span(B)). (More about these
notational conventions can be found in Pollock (1979, chapter 3)). Hence the
geometrical interpretation of (1.2) is formally given by

$$(1.3) \qquad \xi \in O(span(B)).$$

Let a metric on R^k be defined by a symmetric, positive definite (k×k) matrix

Q. It is clear that orthogonality and distances are defined with respect to some inner product definition in a metric space. When x is defined to be a vector in $\mathbb{R}^k$, like ξ, then the inner product in a metric defined by Q is given by

(1.4) $(x,\xi)_Q = x'Q\xi.$

The vectors x and ξ are said to be Q-orthogonal when their inner product equals 0. The squared norm (or length) of ξ in this metric is defined by

(1.5) $\|\xi\|_Q^2 = (\xi,\xi)_Q = \xi'Q\xi.$

The squared norm of x-ξ, given by

(1.6) $\|x-\xi\|_Q^2 = (x-\xi)'Q(x-\xi),$

defines the squared Q-distance between x and ξ, which will be denoted by $d_Q^2(x,\xi)$. The aforementioned metric characteristics are in the following used to described the minimum distance problem.

The problem of deriving a "best fitting" linear subspace for the random vector X can formally be described by means of two successive mathematical programming problems. The first objective is to obtain a vector $\tilde{X} \in O(\text{span}(B))$, hence $B'\tilde{X} = 0$, with minimum squared Q-distance to X. Using ξ as an auxiliary vector this minimum distance problem is given by

(1.7) $\min_{\substack{\xi \\ B'\xi=0}} (X-\xi)'Q(X-\xi).$

The optimal value $\tilde{X}$ of ξ is obtained as a function of X, Q and B. The second objective is then to find the linear subspace $O(\text{span}(B))$, such that the expected squared distance between X and $\tilde{X}$ is minimal. The optimal linear subspace is described by matrix B, which can be obtained by minimizing the expected squared Q-distance between X and $\tilde{X}$ with respect to B. In mathematical terminology this problem is described by

(1.8) $\min_B E(X-\tilde{X})'Q(X-\tilde{X}).$

Expressions (1.7) and (1.8) together yield the composite mathematical programming problem

(1.9) $\min\limits_{\substack{B}} E[\min\limits_{\substack{\xi \\ B'\xi=0}} (X-\xi)'Q(X-\xi)]$.

So far, the minimum distance problem has been defined in the metric defined by Q. By means of a coordinate transformation of X, this problem can be reduced to a minimization of I_k-distances. Before developing such a transformation let us first observe the factorization of the metric-defining matrix Q. As the matrix Q is symmetric, the characteristic vectors of Q can be chosen as an orthonormal set, i.e. there exists a matrix K with characteristic vectors and a diagonal matrix Δ of characteristic roots, such that

(1.10) $QK = K\Delta$ and $K'K = KK' = I_k$.

Denoting the matrix product $K\Delta^{\frac{1}{2}}K'$ (where $\Delta^{\frac{1}{2}}$ stands for the diagonal matrix with the square roots of the charateristic roots on its diagonal) by $Q^{\frac{1}{2}}$, it follows from (1.10) that

(1.11) $Q^{\frac{1}{2}}Q^{\frac{1}{2}} = Q$ and $Q^{\frac{1}{2}}Q^{-1}Q^{\frac{1}{2}} = I_k$.

The notation $Q^{\frac{1}{2}}$ is chosen as this matrix is obviously the generalization of a square root. Also the inverse of $Q^{\frac{1}{2}}$, say $Q^{-\frac{1}{2}}$, exists and is given by

(1.12) $Q^{-\frac{1}{2}} = K\Delta^{-\frac{1}{2}}K'$.

Consider the transformation of vector X given by

(1.13) $X^* = Q^{\frac{1}{2}}X$.

Then the minimization problem denoted in (1.9) can be rewritten in the metric defined by I_k as

(1.14) $\min\limits_{\substack{B^*}} E[\min\limits_{\substack{\xi^* \\ B^{*}{}'\xi^*=0}} (X^*-\xi^*)'(X^*-\xi^*)]$

with $\xi^* = Q^{\frac{1}{2}}\xi$ and $B^* = Q^{-\frac{1}{2}}B$.

Although this preliminary transformation has some definite advantages, which will be used later in this section in order to describe an algorithm that computes the optimal value of B, we abstain from this procedure as it has disadvantages for the structure of this study. In fact, in the sections to come the metric matrix will be used to define alternative forms of the minimum distance problem. Solutions for minimization problem (1.9), the distance minimization problem in its most general form, are given by the following theorem.

Theorem (the PR matrix)
Let the mathematical programming problem be defined by (1.9), with

    X a $(k \times 1)$ random vector
    ξ a $(k \times 1)$ auxiliary vector
    B a $(k \times p)$ parameter matrix of rank p
    Q a $(k \times k)$ positive definite matrix.

The optimal value $\tilde{X}$ of ξ is then given by

$$(1.15) \qquad \tilde{X} = (I_k - Q^{-1}B(B'Q^{-1}B)^{-1}B')X$$

and the optimal value B_{pr}, say, of B is given by

$$(1.16) \qquad \Sigma B_{pr} = Q^{-1}B_{pr}\Lambda_{min,p}$$

with Σ the covariance matrix of X and $\Lambda_{min,p}$ the $(p \times p)$ diagonal matrix with on its diagonal the p smallest characteristic roots of Σ in the metric defined by Q^{-1}.

Before we come to the proof of this theorem, the results presented in (1.15) and (1.16) are considered more closely. Therefore one must realize that from projection theory it is well-known that when A is a $(k \times p)$ matrix of rank p (with $p \leq k$), then $A(A'QA)^{-1}AQ$ is a Q-orthogonal projector onto span(A) and $(I_k - A(A'QA)^{-1}A'Q)$ is a Q-orthogonal projector onto $O_Q(span(A))$, where the subscript Q in connection with O denotes orthogonality in the Euclidean metric

defined by matrix Q. (More on projection theory can, for instance, be found in
Afriat (1957) and Pollock (1979, chapters 2 and 3)). From these facts it
follows directly that $Q^{-1}B(B'Q^{-1}B)^{-1}B'$ in expression (1.15) defines the Q-
orthogonal projector on span($Q^{-1}B$) and that $\tilde{X}$ represents the Q-orthogonal
projection of X on $O_Q(\text{span}(Q^{-1}B))$. Note that the projection is on
$O_Q(\text{span}(Q^{-1}B))$ and not on $O_Q(\text{span}(B))$. This implies that $\tilde{X} \in O_Q(\text{span}Q^{-1}B))$, or
in other words, $\tilde{X}'QQ^{-1}B = \tilde{X}'B = 0$, so that the restriction $B'\tilde{X} = 0$ represents
orthogonality in the I_k-metric, while the projection takes place in the Q-
metric. Formula (1.16) symbolizes the fact that the columns of B_{pr} are the
characteristic vectors of Σ corresponding with the p smallest characteristic
roots in the Q^{-1}-metric. It is evident that the characteristic root λ of Σ in
the metric defined by Q^{-1} is found by solving the polynomial equation
$|\Sigma - \lambda Q^{-1}| = 0$.

Proof

The Lagrangian function of the inner minimization problem (1.7) is given by

(1.17) $L = (X-\xi)'Q(X-\xi) - 2\kappa'B'\xi$

with κ the (p×1) vector of Lagrange multipliers. The optimal values
$\tilde{X}$ and κ_o for ξ and κ of the Lagrangian function are determined by setting the
partial derivatives of L with respect to ξ and κ equal to zero.

(1.18) $\left.\dfrac{\partial L}{\partial \xi}\right|_{\substack{\xi=\tilde{X} \\ \kappa=\kappa_o}} = -2QX + 2Q\tilde{X} - 2B\kappa_o = 0$

(1.19) $\left.\dfrac{\partial L}{\partial \kappa}\right|_{\substack{\xi=\tilde{X} \\ \kappa=\kappa_o}} = -2B'\tilde{X} = 0.$

Rewriting (1.18) as

(1.20) $\tilde{X} = X + Q^{-1}B\kappa_o$

and substituting (1.20) in (1.19), we obtain

(1.21) $\kappa_o = -(B'Q^{-1}B)^{-1}B'X.$

Note that as Q^{-1} is positive definite and B is of rank p, the matrix $B'Q^{-1}B$ is also positive definite and therefore non-singular (see for instance Muirhead (1982, p. 585)). Substitution of (1.21) in (1.20) yields the optimal value $\tilde{X}$ as proposed in (1.15).

Substituting result (1.15) in (1.8), the other minimization problem (with respect to B) can be written as

(1.22) $\min_{B} E(X'B(B'Q^{-1}B)^{-1}B'X).$

Using the trace operator, (1.22) can be rewritten as

(1.23) $\min_{B} tr(B(B'Q^{-1}B)^{-1}B'\Sigma).$

The derivative of (1.23) with respect to B is known in the literature and can for instance be found in Balestra (1976, p. 74). Setting this derivative equal to zero, we obtain the optimal value B_{pr} of B from matrix equation

(1.24) $(I_k - Q^{-1}B_{pr}(B'_{pr}Q^{-1}B_{pr})^{-1}B'_{pr})\Sigma B_{pr}(B'_{pr}Q^{-1}B_{pr})^{-1} = 0$ (k×p).

The resulting system of equations seems a little obscure, as the parameter matrix B_{pr} appears many times in the matrix equation. In order to get an idea of a possible solution, consider $Q = I_k$ and restrict the possible values of B_{pr} to the class of orthogonal matrices such that $B'_{pr}B_{pr} = I_p$. Then (1.23) is the minimum of a quadratic form in B and it is well-known that its solution is given by the set of characteristic vectors of Σ, corresponding with the p smalles characteristic values. Generalizing this approach and assuming that $B'_{pr}Q^{-1}B_{pr} = I_p$, then from (1.24) it follows that

(1.25) $B'_{pr}Q^{-1}B_{pr} = I_p \Rightarrow (I_k - Q^{-1}B_{pr}B'_{pr})\Sigma B_{pr} = 0.$

The right-hand side of (1.25) is known to be solved by the characteristic vectors of Σ in the Q^{-1}-metric, or, more formally, the stationary value of B_{pr} is defined by the system of equations

(1.26) $\Sigma B_{pr} = Q^{-1}B_{pr}\Lambda_p$

with Λ_p a diagonal matrix, with p of the k characteristic roots of Σ in the Q^{-1}-metric on its diagonal.

Let us return to our original equation (1.24). Note that also without the orthogonality assumption (1.26) defines a solution of (1.24), which may be checked by substitution of $Q^{-1}B_{pr}\Lambda_p$ for ΣB_{pr} in (1.24) and verification of the equality sign. However, this leaves us with the question which p characteristic vectors have to be selected. Pre-multiplying (1.26) by B'_{pr} and using the resulting equality together with (1.23) it is deduced that the minimum of $tr(B(B'Q^{-1}B)^{-1}B'\Sigma)$ is given by $tr(\Lambda_p)$. As a result we may conclude that the columns of B_{pr} equal the characteristic vectors of Σ corresponding with the p smallest characteristic roots in the Q^{-1}-metric, which had to be shown. ◊

Four remarks with respect to the solution B_{pr} should be made here:

- Besides the matrix B_{pr} of characteristic vectors, there may be more matrices B that fulfill relation (1.24), viz. any basis of p vectors spanning the same subspace as B_{pr} does. However, matrix B_{pr} as proposed seems to be a natural choice as the characteristic roots and vectors are well-known matrix functions with some nice properties. For instance, the characteristic vectors of a symmetric matrix (like Σ in this case) are known to be orthogonal in the metric defined by Q^{-1}. Therefore the columns of B_{pr} indicate an orthogonal decomposition of span(B_{pr}). (See for instance Pollock (1979, p. 58)).
- The characteristic roots and vectors of Σ in the Q^{-1}-metric correspond with the characteristic roots and vectors of $Q\Sigma$ in the "simple" I_k-metric. The characteristic roots of $Q\Sigma$ are exactly the characteristic roots of ΣQ (see Dhrymes (1978, p. 470)).
- Characteristic vectors are evidently not unique. If c is a non-null scalar, then $c\beta$ is a characteristic vector, if β is. A way of determining one fixed solution may be to normalize the columns of B_{pr} so that their lengths equal 1, or equal the corresponding characteristic roots. Together with the orthogonality property this is in mathematical terminology described by either

(1.27)   unit normalization: $B'_{pr}Q^{-1}B_{pr} = I_p$

or

(1.28)   characteristic root normalization: $B'_{pr}Q^{-1}B_{pr} = \Lambda_{min,p}$.

– Another natural way of normalizing B_{pr} is to set in every column one specific component equal to 1 or -1. For instance PR analysis with $Q = I_k$, $p=1$ and one of the components of (vector) B_{pr} fixed and equal to -1, boils down to orthogonal regression analysis.

Recall the transformation of X, given by $X^* = Q^{\frac{1}{2}}X$, which has been defined in order to transform the minimization problem into the metric defined by I_k. From (1.14) and the results of the preceding theorem it follows that the PR analysis of the transformed vector X^* in the I_k-metric yields solutions $\widetilde{X}^*$ and B_{PR}^* that are given by

$$(1.29) \qquad \widetilde{X}^* = (I_k - B^*(B^{*'}B^*)^{-1}B^{*'})X^*$$

and

$$(1.30) \qquad \Sigma^* B_{pr}^* = B_{pr}^* \Lambda_{min,p}^*.$$

with Σ^* the covariance matrix of X^* given by

$$(1.31) \qquad \Sigma^* = E(X^*X^{*'}) = Q^{\frac{1}{2}}\Sigma Q^{\frac{1}{2}}.$$

Combining (1.30) and the untransformed counterpart $\Sigma B_{pr} = Q^{-1}B_{pr}\Lambda_{min,p}$, and recalling that $B_{pr}^* = Q^{-\frac{1}{2}}B_{pr}$ it follows that

$$(1.32) \qquad \Lambda_{min,p}^* = \Lambda_{min,p}.$$

Note that normalizing the columns of B_{pr}^* in the I_k-metric is equivalent to normalizing the columns of B_{pr} in the Q^{-1}-metric, as $B_{pr}^{*'}B_{pr}^* = B_{pr}'Q^{-1}B_{pr}$. More propositions about projections of R^k on linear subspaces in the I_k-metric, that are obtained as extrema of quadratic forms, can be found in Rao (1973, section 1f). The advantage of transformation $X^* = Q^{\frac{1}{2}}X$ is that the parameters are obtained as characteristic vectors of a symmetric matrix in the simple Euclidean metric defined by I_k. The procedure of deriving characteristic vectors is included in many software packages, hence once an estimator of Σ has been obtained, the computation of the parameter matrix B_{pr}^* will be no problem.

In what follows we shall proceed from an asymptotically normally distributed estimator $\hat{\Sigma}$ of covariance matrix Σ. In section II.2 it has been shown that this requirement is fulfilled when $\hat{\Sigma}$ equals the sample covariance matrix of a random sample consisting of N i.i.d. observations on X. Furthermore, the method of moments was proposed for parameters that are defined as continuous functions of Σ (or in fact of vecΣ), as it yields consistent estimators. Here the parameter vectors of interest are given as transformations of the characteristic vectors of Σ. Although the characteristic vectors cannot be obtained as an explicit function of Σ, the method of moments can be applied as they are implicitly defined as functions of Σ. Basically the MME is obtained by applying the results of the preceding theorems to $\hat{\Sigma}$. We have now the following algorithm to compute the MME $\hat{B}_{pr}$ of B_{pr}:

1. compute $\hat{\Sigma}$ (an asymptotically normal estimator of Σ);
2. deduce from Q its characteristic roots and vectors (Δ and K);
3. $Q^{\frac{1}{2}} := K\Delta^{\frac{1}{2}}K'$;
4. $\hat{\Sigma}^* := Q^{\frac{1}{2}}\hat{\Sigma}Q^{\frac{1}{2}}$;
5. deduce from $\hat{\Sigma}^*$ its p minimal characteristic roots and vectors ($\hat{\Lambda}_{min,p}$ and $\hat{B}_{pr}^*$);
6. $\hat{B}_{pr} := Q^{\frac{1}{2}}\hat{B}_{pr}^*$.

As $\hat{B}_{pr}$ is a function of the random matrix $\hat{\Sigma}$ it is a random matrix itself and we are interested in its limiting distribution. As the vectorized version of the estimated matrix (vec$\hat{B}_{pr}$) can be considered as a function of the vectorized form of $\hat{\Sigma}$, it follows from the asymptotic normality of vec$\hat{\Sigma}$ that vec$\hat{B}_{pr}$ is also asymptotically normally distributed. The expectation of the limiting distribution of vec$\hat{B}_{pr}$ is given by vecB_{pr} and its covariance matrix can be obtained by applying the delta method. Therefore the derivatives of vecB_{pr} with respect to (vecΣ)' are needed. Defining the (k×1) vector β_i as th ith column of B_{pr}, hence

(1.33) $B_{pr} = [\beta_1,\ldots,\beta_p]$,

the matrix of partial derivatives is given by

$$(1.34) \qquad \nabla(B_{pr};\Sigma) = \frac{\partial(\text{vec}B_{pr})}{\partial(\text{vec}\Sigma)'} = \begin{bmatrix} \dfrac{\partial\beta_1}{\partial(\text{vec}\Sigma)'} \\ \vdots \\ \dfrac{\partial\beta_p}{\partial(\text{vec}\Sigma)'} \end{bmatrix}. \qquad (kp \times k^2)$$

Obviously it is sufficient to obtain the derivative of one column of B_{pr} (the derivative of a characteristic vector of Σ in the Q^{-1}-metric) with respect to $(\text{vec}\Sigma)'$.

Theorem (the derivatives of the PR matrix)

Let λ and β_{pr} be a characteristic root and the corresponding unique characteristic vector of Σ in the Q^{-1}-metric, that is

$$(1.35) \qquad \Sigma\beta_{pr} = Q^{-1}\beta_{pr}\lambda$$

with normalization

$$(1.36) \qquad \beta'_{pr}Q^{-1}\beta_{pr} = 1.$$

Then β_{pr} may be considered as a differentiable function of Σ with

$$(1.37) \qquad \nabla(\beta_{pr};\Sigma) = -(\beta'_{pr} \otimes Q^{\frac{1}{2}}(\Sigma^* - \lambda I_k)^+ Q^{\frac{1}{2}}) \qquad (k \times k^2)$$

with $Q^{\frac{1}{2}}$ the transformation matrix defined in (1.11), $\Sigma^* = Q^{\frac{1}{2}}\Sigma Q^{\frac{1}{2}}$ and $(\Sigma^* - \lambda I_k)^+$ the Moore–Penrose inverse of $(\Sigma^* - \lambda I_k)$.

Proof

Combining (1.35) and (1.36) it follows that

$$(1.38) \qquad \beta'_{pr}\Sigma\beta_{pr} = \lambda.$$

Using (1.38), expression (1.35) can be written as

$$(1.39) \qquad (I_k - Q^{-1}\beta_{pr}\beta'_{pr})\Sigma\beta_{pr} = 0.$$

From (1.39) we deduce by taking the total differential with respect to $(\mathrm{vec}\Sigma)'$

$$(1.40) \quad \nabla((I_k - Q^{-1}\beta_{pr}\beta'_{pr})\Sigma\beta_{pr};\Sigma) + \nabla((I_k - Q^{-1}\beta_{pr}\beta'_{pr})\Sigma\beta_{pr};\beta_{pr}) \ \nabla(\beta_{pr};\Sigma) = 0.$$

Note that we are interested in the second factor of the second term of equation (1.40). An expression for this gradient matrix will be obtained by first analysing the two other matrix derivatives in (1.40). Applying matrix differentiation properties (A.46) and (A.42), the first term of (1.40) can be rewritten as

$$(1.41) \quad (I_k \otimes (\mathrm{vec}I_k)')((\mathrm{vec}(I_k - Q^{-1}\beta_{pr}\beta'_{pr}))(\beta'_{pr} \otimes I_k).$$

Using (A.11) and (A.13) we obtain the following result

$$(1.42) \quad \nabla((I_k - Q^{-1}\beta_{pr}\beta'_{pr})\Sigma\beta_{pr};\Sigma) = (\beta'_{pr} \otimes (I_k - Q^{-1}\beta_{pr}\beta'_{pr})).$$

Application of (A.41), (A.44) and (A.39) to the first factor of the second term of (1.40) yields

$$(1.43) \quad -Q^{-1}((\beta'_{pr} \otimes I_k) + \beta_{pr}(\mathrm{vec}I_k)')(\Sigma\beta_{pr} \otimes I_k) + (I_k - Q^{-1}\beta_{pr}\beta'_{pr})\Sigma.$$

Using some Kronecker product rules and equality (1.38) we get

$$(1.44) \quad \nabla((I_k - Q^{-1}\beta_{pr}\beta'_{pr})\Sigma\beta_{pr};\beta_{pr}) = (\Sigma-\lambda Q^{-1}) - 2Q^{-1}\beta_{pr}\beta'_{pr}\Sigma.$$

Substituting (1.42) and (1.44) in (1.40) we obtain

$$(1.45) \quad (\beta'_{pr} \otimes (I_k - \beta_{pr}\beta'_{pr})) + ((\Sigma-\lambda Q^{-1}) - 2Q^{-1}\beta_{pr}\beta'_{pr}\Sigma) \ \nabla(\beta_{pr};\Sigma) = 0$$

which can easily be rewritten as

$$(1.46) \quad (\Sigma-\lambda Q^{-1}) \ \nabla(\beta_{pr};\Sigma) =$$

$$= -(\beta'_{pr} \otimes I_k) + (\beta'_{pr} \otimes Q^{-1}\beta_{pr}\beta'_{pr}) + 2Q^{-1}\beta_{pr}\beta'_{pr}\Sigma \ \nabla(\beta_{pr};\Sigma).$$

Substitution of $Q^{-\frac{1}{2}}\Sigma^*Q^{-\frac{1}{2}}$ for Σ yields for the first factor at the left-hand side of (1.46)

(1.47) $(\Sigma - \lambda Q^{-1}) = Q^{-\frac{1}{2}}(\Sigma^* - \lambda I_k)Q^{-\frac{1}{2}}.$

Using the Moore-Penrose generalized inverse of $(\Sigma^* - \lambda I_k)$, (note that $(\Sigma^* - \lambda I_k)$ is singular as $|\Sigma^* - \lambda I_k| = 0$), (1.46) can by means of (1.47) be rewritten as

(1.48) $V(\beta_{pr}; \Sigma) = -(\beta'_{pr} \otimes Q^{\frac{1}{2}}(\Sigma^* - \lambda I_k)^+ Q^{\frac{1}{2}})$

$$+ (\beta'_{pr} \otimes Q^{\frac{1}{2}}(\Sigma^* - \lambda I_k)^+ Q^{-\frac{1}{2}} \beta_{pr} \beta'_{pr})$$

$$+ 2Q^{\frac{1}{2}}(\Sigma^* - \lambda I_k)^+ Q^{-\frac{1}{2}} \beta_{pr} \beta'_{pr} \Sigma \, V(\beta_{pr}; \Sigma)$$

$$+ Q^{\frac{1}{2}}(I_k - (\Sigma^* - \lambda I_k)^+(\Sigma^* - \lambda I_k))U$$

with U an arbitrary $(k \times k^2)$ matrix. The second and the third term at the right-hand side of (1.48) both contain a factor $(\Sigma^* - \lambda I_k)^+ Q^{-\frac{1}{2}} \beta_{pr} = (\Sigma^* - \lambda I_k)^+ \beta^*_{pr}$. As $(\Sigma^* - \lambda I_k)\beta^*_{pr} = 0$, it follows from (A.25) and the symmetry of $(\Sigma^* - \lambda I_k)$ that

(1.49) $(\Sigma^* - \lambda I_k)^+ Q^{-\frac{1}{2}} \beta_{pr} = 0.$

Because of the unit normalization (1.36) we have

(1.50) $\beta'_{pr} Q^{-1} V(\beta_{pr}; \Sigma) = 0,$

and combination of (1.48), (1.49) and (1.50) yields

(1.51) $\beta'_{pr} Q^{-\frac{1}{2}} U = \beta^*_{pr}{}' U = 0.$

Moreover, the uniqueness of the characteristic vector β_{pr} yields

(1.52) $I_k - (\Sigma^* - \lambda I_k)^+(\Sigma^* - \lambda I_k)U = \beta^*_{pr} \beta^*_{pr}{}' U$

which equals zero because of (1.51). From substitution of (1.49) and (1.52) in (1.48) it follows that the last three terms at the right-hand side of (1.48) equal zero, which completes the proof. $\diamond$

Let us briefly consider the formula of the derivative of the PR matrix that has been obtained. First of all one should note that the transformation matrix $Q^{\frac{1}{2}}$ has been used to find an attractive expression for the gradient matrix $\nabla(\beta_{pr};\Sigma)$. Acting this way, two of the four terms in (1.48) can be shown to equal zero. As $Q^{\frac{1}{2}}(\Sigma^{*}-\lambda I_{k})^{+}Q^{\frac{1}{2}}$ does not necessarily equal $(Q^{-\frac{1}{2}}\Sigma^{*}Q^{-\frac{1}{2}}-\lambda Q^{-\frac{1}{2}}Q^{-\frac{1}{2}}) = (\Sigma-\lambda Q^{-1})^{+}$, the matrix $Q^{\frac{1}{2}}$ remains in the final expression for $\nabla(\beta_{pr};\Sigma)$.
The uniqueness of the generalized inverse in the gradient matrix is guaranteed by the uniqueness of the characteristic vector that corresponds with the characteristic root λ. It is for this reason that all the characteristic roots that will be dealt with in this study, are assumed to be distinct (see also figure V.2.1).
Similar formulas have been derived in the context of Linear Operator Analysis. In this theory, a matrix is observed as a linear operator for which a convergent power series exist. Together with these series, one observes the power series of the corresponding characteristic roots and vectors. A frequently used procedure dealing with convergent power series is called "analytic perturbation theory" and has for instance been studied by Kato (1955, 1966) and Friedrichs (1965). This method uses perturbed (or disturbed) operators, yielding differential equations with characteristic roots and vectors. Essentially, there is not much difference between the perturbation method and the approach applied here, that employs the tools of differential calculus directly. The perturbation method is intuitively somewhat clearer as it is based on the definition of a derivative as a ratio of infinitely small differences.

Combination of the results for $\nabla(\beta_{pr};\Sigma)$ with the delta method yields the following limiting distribution of $\text{vec}\hat{B}_{pr}$

$$(1.53) \qquad \sqrt{N}(\text{vec}\hat{B}_{pr} - \text{vec}B_{pr}) \xrightarrow{D} N(0, [\nabla(\text{vec}B_{pr};\Sigma)]V[\nabla(\text{vec}B_{pr};\Sigma)]')$$

with N the number of observations in the sample and V the asymptotic covariance matrix of $\sqrt{N}(\text{vec}\hat{\Sigma} - \text{vec}\Sigma)$.

In section I.2. the Population-Sample Decomposition (PSD) principle was explained. Note the convenience of this technique in deriving the limiting

distribution of $\text{vec}\hat{B}_{pr}$. We only need to define the parameters as a
differentiable (implicit) function of Σ with its corresponding gradient. This
was called the population part of the problem, as nothing about the
distributional form of the sample observations is assumed. The limiting
distribution of $\text{vec}\,\hat{B}_{pr}$ is "automatically" provided by the asymptotic normality
of $\text{vec}\hat{\Sigma}$. Using the PSD approach in case of N random i.i.d. observations of X,
an explicit specification of the matrix V can be obtained. Performing the pre-
and post-multiplication of V by the matrix of partial derivatives and its
transpose, a simple form of the covariance matrix of the limiting distribution
as denoted in (1.53) is obtained. In the following theorem the results for a
random i.i.d. sample will be given for the simple case of a parameter vector
β_{pr} (hence p=1). These results can easily be extended to the more general case
of several vectors, yielding a parameter matrix B_{pr} (hence $1 < p < k$).

Theorem ($\hat{\beta}_{PR}$ for a random i.i.d. sample)
Let $\{X_n: n=1,\ldots,N\}$ be a sample of N i.i.d. random observations on X and let
$\hat{\beta}_{pr}$ be the MME of β_{pr}, that is defined as the characteristic vector of Σ in
the metric defined by Q^{-1} (see (1.35) and (1.36)). Then

$$(1.54) \qquad \sqrt{N}(\hat{\beta}_{pr} - \beta_{pr}) \xrightarrow{D} N(0, \; Q^{\frac{1}{2}}(\Sigma^*-\lambda I_k)^+Q^{\frac{1}{2}}[E(XX'\beta_{pr}\beta'_{pr}XX')]Q^{\frac{1}{2}}(\Sigma^*-\lambda I_k)^+Q^{\frac{1}{2}})$$

with λ the characteristic root corresponding with β_{pr}, $Q^{\frac{1}{2}}$ the transformation
matrix defined in (1.11) and $\Sigma^* = Q^{\frac{1}{2}}\Sigma Q^{\frac{1}{2}}$.

Proof
Combining the formula of the limiting distribution given in (1.53) with the
gradient defined in (1.37) it follows that $\sqrt{N}(\hat{\beta}_{pr} - \beta_{pr})$ is asymptotically
normally distributed with mean zero and covariance matrix

$$(1.55) \qquad (\beta'_{pr} \otimes Q^{\frac{1}{2}}(\Sigma^*-\lambda I_k)^+Q^{\frac{1}{2}})V(\beta_{pr} \otimes Q^{\frac{1}{2}}(\Sigma^*-\lambda I_k)^+Q^{\frac{1}{2}}).$$

As the sample observations are i.i.d., the matrix V equals the covariance
matrix of $\text{vec}(XX')$, e.g.

$$(1.56) \qquad V = E(\text{vec}(XX'))(\text{vec}(XX'))' - (\text{vec}\Sigma)(\text{vec}\Sigma)'.$$

(Recall that the expectation of X was assumed to be zero and see also the theorem on asymptotic normality of vec$\hat{\Sigma}$ in section II.2). Substituting (1.56) in (1.55) and applying (A.11) four times, the vec-operators can be eliminated and (1.55) results in

$$(1.57) \qquad Q^{\frac{1}{2}}(\Sigma^{*}-\lambda I_{k})^{+}Q^{\frac{1}{2}}[E(XX'\beta_{pr}\beta'_{pr}XX') - \Sigma\beta_{pr}\beta'_{pr}\Sigma]Q^{\frac{1}{2}}(\Sigma^{*}-\lambda I_{k})^{+}Q^{\frac{1}{2}}.$$

Replacing $\Sigma\beta_{pr}$ by $Q^{-1}\beta_{pr}\lambda$ and using (1.49) it follows that (1.57) equals the limiting covariance matrix proposed in (1.54). ◊

In this section we studied the dimension reduction of $(k \times 1)$ vector X in a linear vector space with metric defined by Q. Projection of X on a linear subspace yielded a vector $\tilde{X}$ in a $(k-p)$-dimensional subspace with minimum Q-distance to X, such that $B'\tilde{X} = 0$. Replacing the random vector $\tilde{X}$ by N random observations X_n on X, we obtain a relation that stands for a linear model specification for observation X_n $(n=1,\ldots,N)$. Comparing this specification with well-known model specifications in statistical (especially econometric, biometric and psychometric) literature it should be noticed that this model does not include any so-called exogenous (or fixed) variables. All components of X are random. Therefore $B'X_n = 0$ may be considered as a structural linear model including endogenous variables only. In order to relate the minimum distance approach with statistical linear relations that include exogenous as well as endogenous variables (like simultaneous equations systems and linear regression equations), the relations between the definitions of endogeneity, exogeneity and the projection direction of X need to be specified.

III.2 The distance matrix Q

In the preceding section the parameter matrix B has been obtained as a function of covariance matrix Σ by minimizing the (expected squared) distance between X and ξ under the restriction $B'\xi = 0$. The distance was assumed to be an Euclidean one, defined by a positive definite matrix Q. The geometrical interpretation of this mathematical programming problem is that X is projected onto the linear subspace given by $\{\xi: B'\xi=0\}$, yielding a transformation of X given by

$$(2.1) \qquad \tilde{X} = (I_k - Q^{-1}B(B'Q^{-1}B)^{-1}B')X.$$

It has been noted that $\tilde{X}$ is a Q-orthogonal projection of X on the linear subspace given by $O_Q(\text{span}(Q^{-1}B))$. The choice of Q influences the projection direction and consequently the way in which X is transformed in order to obtain $\tilde{X}$. We shall start by elucidating how the various components of Q affect the projection of X.

Let it first be supposed that Q is a diagonal matrix, that is

$$(2.2) \qquad Q_d = \text{diag}(q_1,\ldots,q_k) \qquad\qquad \text{with } q_1,\ldots,q_k > 0.$$

Geometrically this means that the distance minimization problem is observed in a vector space spanned by a set of orthogonal coordinate axes, where the distances along the k axes are measured in different units. By applying a coordinate transformation one can obtain a basis such that the distances along the coordinate axes are measured in the same units. Indicating the transformation of X by X^*, the transformation is given by $X_i^* = \sqrt{q_i}\cdot X_i$, or in matrix notation, the change of basis is obtained by pre-multiplication of X by $Q_d^{\frac{1}{2}} = \text{diag}(\sqrt{q_1},\ldots,\sqrt{q_k})$.

In general the distance defining matrix is not necessarily diagonal. Positive definiteness of Q is sufficient to let $d_Q(.,.)$ define a distance function. A distance defined by a non-diagonal matrix Q does not only measure the distances along the axes in different units, but it also implies that the coordinate axes are no longer orthogonal. The distances are measured in a space, spanned by a set of oblique coordinate axes. In section 1 it has been indicated how to obtain the corresponding distances in a vector space, spanned by an orthonormal basis, by applying a coordinate transformation. It was suggested to pre-multiply X by matrix $Q^{\frac{1}{2}}$, which is defined as the (k×k) symmetric matrix that satisfies $Q^{\frac{1}{2}}Q^{\frac{1}{2}} = Q$. Note that this definition of $Q^{\frac{1}{2}}$ is a generalization of $Q_d^{\frac{1}{2}}$, which has been defined for a diagonal matrix Q.

How can this information be used to choose an appropriate matrix Q? It is known that $(X-\tilde{X})$ must be seen in a probabilistic context and that the partial deviations $(X_i-\tilde{X}_i)$ may have different variances. Intuitively, it seems logical that the deviation of $\tilde{X}_i$ from X_i should carry a high weight in the minimization problem when its variance is small and, on the contrary, that a partial deviation with a large variance should carry a low weight in the determination of the minimum distance between X and $\tilde{X}$. In case of a diagonal matrix Q this may be formalized by setting the weights q_i equal to the inverse of the variance of $(X_i-\tilde{X}_i)$. This idea is generalized for a non-diagonal matrix Q by observing the projection distances of $(X-\tilde{X})$ in a metric defined by the inverse of $E(X-\tilde{X})(X-\tilde{X})'$. Hence defining a metric matrix Q_* by

$$(2.3) \qquad Q_* = (E(X-\tilde{X})(X-\tilde{X})')^{-1},$$

we create a new scale in which the components of $(X-\tilde{X})$ with high (co)variances cause a large freedom of variability in the corresponding dimension of the projection. Observing $(X-\tilde{X})$ in a metric defined by Q_*, corresponds with an examination of the transformed vector $(X^*-\tilde{X}^*) = Q_*^{\frac{1}{2}}(X-\tilde{X})$ in the simple Euclidean metric defined by I_k, with $Q_*^{\frac{1}{2}}$ such that $Q_*^{\frac{1}{2}}Q_*^{\frac{1}{2}} = Q_*$ and $Q_*^{\frac{1}{2}}Q_*^{-1}Q_*^{\frac{1}{2}} = I_k$. This means that the partial deviations $(X_i^*-\tilde{X}_i^*)$ for i = 1,...,k, are all equally weighted in the distance minimization problem. This harmonizes with the idea of connecting high weights with small partial deviation variances, and the covariance matrix of $(X^*-\tilde{X}^*)$ is given by

$$(2.4) \qquad E(X^*-\tilde{X}^*)(X^*-\tilde{X}^*)' = Q_*^{\frac{1}{2}}(E(X-\tilde{X})(X-\tilde{X})')Q_*^{\frac{1}{2}} = Q_*^{\frac{1}{2}}Q_*^{-1}Q_*^{\frac{1}{2}} = I_k.$$

When the minimum distance parameters are estimated in the metric defined by Q_*, we obtain an iterative estimation problem, as Q_* is determined by the distribution of $(X-\tilde{X})$. In order to estimate the parameter matrix B_{pr}, the matrix $Q_* = (E(X-\tilde{X})(X-\tilde{X})')^{-1}$ is needed, but we can only obtain approximations of $\tilde{X}$, and therefore estimators of Q_*, when we have knowledge of the values of B_{pr}. Distribution of these type of estimators will here be left out of consideration and in the sections to come the study will be concentrated on some specific predetermined choices for Q.

The choice of the metric co-determines the form of the optimal value of parameter matrix B, as B is obtained as a function of Σ and Q. Particularly when some components of Q attain extreme values like 0 or ∞, the analyses of minimum distance parameters are radically influenced. Let us observe the geometrical aspects of some specific metrics defined by matrices with 0 and ∞ components. In the limiting case of characteristic roots of Q that tend to infinity, the researcher assigns a very high importance to the exact approximation of population parameters in some dimension of X. For instance when Q is a diagonal matrix, its diagonal elements equal the characteristic roots and the infinity of the ith diagonal element causes the ith dimension of X to be unchanged under a projection in the metric defined by Q, i.e. $\tilde{X}_1 = X_1$. Figures III.2.1, III.2.2 and III.2.3 show the projections of X on the linear subspace defined by $\{\xi: B'\xi=0\}$ with p=1, in a metric defined by

$$I_2 = \begin{bmatrix} 1 & 0 \\ 0 & 1 \end{bmatrix}, \quad \lim_{q_{11} \to \infty} \begin{bmatrix} q_{11} & 0 \\ 0 & 1 \end{bmatrix} = \begin{bmatrix} \infty & 0 \\ 0 & 1 \end{bmatrix} \text{ and } \lim_{q_{22} \to \infty} \begin{bmatrix} 1 & 0 \\ 0 & q_{22} \end{bmatrix} = \begin{bmatrix} 1 & 0 \\ 0 & \infty \end{bmatrix}$$

respectively.

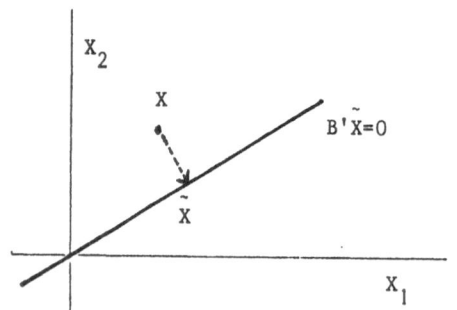

Figure III.2.1 Projection in metric defined by $\begin{bmatrix} 1 & 0 \\ 0 & 1 \end{bmatrix}$.

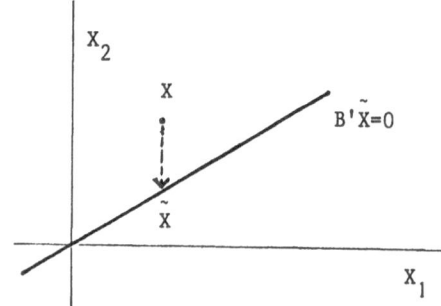

Figure III.2.2 Projection in metric defined by $\begin{bmatrix} \infty & 0 \\ 0 & 1 \end{bmatrix}$.

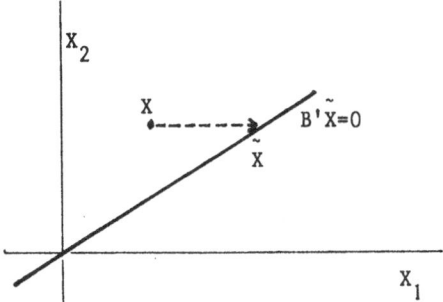

Figure III.2.3 Projection in metric defined by $\begin{bmatrix} 1 & 0 \\ 0 & \infty \end{bmatrix}$.

Figure III.2.2 can be considered as the population analogue of the well-known linear regression model. In functional linear regression analysis we distinguish between two types of variables, viz. the endogenous variables and the exogenous variables. This distinction is not only made in linear regression models, it can also be found in more general (econometric) models, like for instance in simultaneous equations systems. The exogenous variables are used to predict the values of the endogenous variables by means of functional relationships. Therefore endogeneity is often associated with parts of observations that have to be explained and models are based on causality concepts. We shall not dwell on the philosophical interpretations of exogeneity and endogeneity (see e.g. Simon (1979), Wold (1964)), but we shall only consider the implications of the choice of a specific infinite distance defining matrix Q for the definition of linear structural models. Clearly the infinity of the diagonal elements of Q yields a projection of the vector X, that reminds us of linear models in which we distinguish endogenous and exogenous variables. For instance, in case

of a metric defined by $\begin{bmatrix} \infty & 0 \\ 0 & 1 \end{bmatrix}$, like in figure III.2.2, $X_1 = \tilde{X}_1$ behaves as a

fixed component in the linear relation defined by $B'\tilde{X}=0$. In what follows, exogeneity will be defined as fixedness in the projection of X on $\{\xi: B'\xi=0\}$. That is, the components of X that correspond with infinite diagonal elements of Q will be called exogenous variables. In the remaining sections of this

chapter the limiting minimum distance parameter matrix B will be obtained,
when the problem is analysed in a metric defined by a matrix, of which some of
its diagonal elements tend to infinity. Or, in other words, when some of the
components of X are assumed to be exogenous.

Other extreme results are obtained by setting some elements of Q equal to
zero. One should be aware of the fact that this may yield a singular matrix Q.
A singular Q no longer defines a distance function, as one or more of its
characteristic roots equal zero. In a "metric" defined by a positive semi-
definite singular matrix there are sets of different points that have a zero-
distance towards each other. These type of "metrics" are called semi-metrics.
For instance, when Q is a diagonal matrix and its ith diagonal element equals
zero, the semi-metric defined by Q causes the ith dimension of X to have
unlimited freedom. The deviation of $\tilde{X}_i$ from X_i does not influence the distance
between $\tilde{X}$ and X. As a consequence, the ith dimension does not play any role in
the distance minimization analysis and can be neglected. Geometrically, the
omission of the ith component boils down to projecting the vector X, of which
the ith component is omitted, onto a $(k-p-1)$ dimensional subspace in a metric
defined by Q, of which the ith column and i^{th} row are omitted.

Far more interesting are off-diagonal zero's in the matrix Q. For instance
when Q is partitioned as

$$(2.5) \qquad Q = \left[\begin{array}{c:c} Q_1 & 0 \\ \hdashline 0 & Q_2 \end{array} \right]$$

with two off-diagonal blocks of zeros and Q_1 and Q_2 of full rank, then the
linear space R^k in which X and ξ are situated can be considered as the direct
sum of two linear subspaces that are orthogonal in the Euclidean metric
defined by I_k. Thus the vector $(X-\xi)$ consists of two components $(X_1-\xi_1)$ and
$-\xi_2)$ for which the Pythagorean relationship

$$6) \qquad \|X-\xi\|_Q^2 = \|X_1-\xi_1\|_{Q_1}^2 + \|X_2-\xi_2\|_{Q_2}^2$$

lds. Using special forms of the metric, we obtain particular forms of PR
analysis that correspond with well-known multivariate statistical techniques.
Starting in the next section with an analogue of the simultaneous equations
systems in the PSD setting, this analysis will in other parts of this chapter

be specified for special cases like seemingly unrelated regressions and multivariate linear regression theory. It will be seen that PR analysis coincides with canonical correlation analysis when Q is defined to be structured analogous to (2.5), with Q_1 and Q_2 equal to the inverses of submatrices of the covariance matrix Σ.

III.3 Simultaneous equations systems

This section starts by repeating the main mathematical programming problem of interest in the PR analysis (see (1.9))

$$
(3.1) \qquad \min_{B} \; E[\; \min_{\xi} \; (X-\xi)'Q(X-\xi)].
$$
$$
B'\xi=0
$$

Let the vector X be composed of two types of variables that we shall call endogenous and exogenous components of X. In conformity with the definitions of exogeneity and endogeneity in section 2, we put on the exogenous variables the restriction that they remain unchanged under the projection of X. Reorganizing the vector X, the endogenous variables are given by the first m, say, components of X and the last (k-m) components constitute the exogenous part of X. The partitioning is indicated by

$$
(3.2) \qquad X = \begin{bmatrix} Y \\ Z \end{bmatrix}.
$$

with Y the (m×1) endogenous part of X and Z the ((k-m)×1) exogenous (or fixed) part of X. The auxiliary vector ξ, the parameter matrix B, the metric matrix Q and the covariance matrix Σ are partitioned accordingly as

$$
(3.3) \qquad \xi = \begin{bmatrix} \eta \\ \zeta \end{bmatrix}, \quad B = \begin{bmatrix} B_y \\ \hline B_z \end{bmatrix}, \quad Q = \begin{bmatrix} Q_{yy} & Q'_{zy} \\ \hline Q_{zy} & Q_{zz} \end{bmatrix}, \quad \Sigma = \begin{bmatrix} \Sigma_{yy} & \Sigma'_{zy} \\ \hline \Sigma_{zy} & \Sigma_{zz} \end{bmatrix}
$$

The dimensions of the sub-matrices and sub-vectors in (3.3) follow directly from the dimensions of Y and Z. The fact that the exogenous parts of X remain unchanged in the optimal value $\tilde{X}$ can be indicated by

$$
(3.4) \qquad \tilde{X} = \begin{bmatrix} \tilde{Y} \\ \tilde{Z} \end{bmatrix} \quad \text{with } \tilde{Z} = Z , \quad \text{or } \tilde{X} = \begin{bmatrix} \tilde{Y} \\ Z \end{bmatrix}.
$$

The non-variability of Z is included in minimization problem (3.1) by defining Q_{zz} to be a diagonal matrix, the diagonal elements of which tend to infinity, which can be indicated by

$$(3.5) \qquad Q_{zz} = \lim_{q \to \infty} q I_{k-m} = \begin{bmatrix} \infty & & \\ & \infty & \\ & & \ddots & \\ & & & \infty \end{bmatrix} \qquad ((k-m) \times (k-m)).$$

A second assumption we want to make for the projection direction is that it is orthogonal to the linear subspace given by Y = 0. In other words, X is projected parallel to Z = 0. This property is ensured by setting the elements of Q_{zy} equal to zero. Together with (3.5) this yields for the metric matrix Q

$$(3.6) \qquad Q = \left[\begin{array}{c|ccc} Q_{YY} & & 0 & \\ \hline & \infty & & \\ 0 & & \ddots & \\ & & & \infty \end{array} \right] \qquad (k \times k).$$

Substitution of (3.6) in (3.1) yields a mathematical programming problem, the optimal solution of which is obtained by solving the so-called Simultaneous Equations (SE)

$$(3.7) \qquad B'_{se} \tilde{X} = B'_{se,y} \tilde{Y} + B_{se,z} Z = 0.$$

Naturally we could generate the solution, using the results obtained in section 1, by replacing the infinite diagonal elements of Q_{zz} by "very large" values. However, this would yield tricky applications of matrix operations (like matrix multiplication, inversion and deriving characteristic roots and vectors) on matrices with elements that tend to infinity. It seems more appropriate to obtain first the limiting analytical solutions of (3.1) with infinite Q given in (3.6), before assessing the optimal matrix B_{se}. Therefore the problem is rewritten, nòt by substituting an infinite diagonal matrix for Q_{zz}, but by replacing ζ by Z. We obtain

$$(3.8) \qquad \min_{B} E[\min_{\substack{\eta \\ B'_y \eta + B'_z Z = 0}} (Y-\eta)' Q_{yy} (Y-\eta)]$$

with η the remaining variable part of the auxiliary vector ξ. Note that $\tilde{Z} = Z$ requires that $\tilde{Z}$ generates the same subspace as Z, i.e. if Z is a vector in $(k-m)$ dimensions, then $\tilde{Z}$ generates a $(k-m)$-dimensional subspace. It follows that X can only be projected on a subspace of at least dimension $(k-m)$. Hence when the projection space is given by p relations $B_y'\tilde{Y} + B_z'Z = 0$, this means that $k-p \geq k-m$, or $p \leq m$. In case $m=p$, $\tilde{Y}$ can only be a linear combination of Z and as B_y is a square $(p \times p)$ matrix of rank p (recall that B was assumed to be of full column rank) this linear combination is given by

$$(3.9) \qquad \tilde{Y} = -(B_y')^{-1}B_z'Z.$$

This fact simplifies the distance minimization problem drastically, so that the inner minimization vanishes and (3.8) is for $m=p$ rewritten as

$$(3.10) \qquad \min_{B} E[(Y + (B_z B_y^{-1})'Z)'Q_{yy}(Y + (B_z B_y^{-1})'Z)].$$

Taking expectations and using trace properties we obtain

$$(3.11) \qquad \min_{B} [Q_{yy}(\Sigma_{yy} + (B_z B_y^{-1})'\Sigma_{zy} + \Sigma_{zy}'(B_z B_y^{-1}) + (B_z B_y^{-1})'\Sigma_{yy}(B_z B_y^{-1}))],$$

yielding an identification problem for the submatrices B_y and B_z. It is possible to derive an optimal value for matrix product $B_z B_y^{-1}$, but given this solution, it is not possible to obtain optimal values for B_y and B_z separately when no additional information on B_y and/or B_z is available. Before something is said about the identification problem (see also Judge et al. (1982, chapter 12)) and the correspondence between the minimization problem given in (3.8) and the simultaneous equations system theory as known in statistical (and especially econometric) literature, the solution of (3.8) for $p < m$ will be given in the following theorem.

Theorem (the SE matrix)
Let the mathematical programming problem be defined by (3.8) with

Y an $(m \times 1)$ random vector

Z a $((k-m) \times 1)$ random vector

η an $(m \times 1)$ auxiliary vector

$B = [B_y' \mid B_z']'$ a $(k \times p)$ parameter matrix

B_y an $(m \times p)$ matrix of rank p with $p < m$

Q_{yy} an $(m \times m)$ positive definite matrix.

The optimal value $\tilde{Y}$ of η is given by

$$(3.12) \qquad \tilde{Y} = Y - Q_{yy}^{-1} B_y (B_y' Q_{yy}^{-1} B_y)^{-1} B'X$$

and the optimal value B_{se} of B is defined by

$$(3.13) \qquad B_{se} = \begin{bmatrix} B_{se,y} \\ \hline B_{se,z} \end{bmatrix}$$

with $B_{se,z} = -\Sigma_{zz}^{-1} \Sigma_{zy} B_{se,y}$

and $(\Sigma_{yy} - \Sigma_{zy}' \Sigma_{zz}^{-1} \Sigma_{zy}) B_{se,y} = Q_{yy}^{-1} B_{se,y} \Omega_{min,p}$

where $\Sigma = \begin{bmatrix} \Sigma_{yy} & \mid & \Sigma_{zy}' \\ \hline \Sigma_{zy} & \mid & \Sigma_{zz} \end{bmatrix}$ is the covariance matrix of $\begin{bmatrix} Y \\ Z \end{bmatrix}$ and $\Omega_{min,p}$ is the $(p \times p)$

diagonal matrix with on its diagonal the p smallest characteristic roots
of $(\Sigma_{yy} - \Sigma_{zy}' \Sigma_{zz}^{-1} \Sigma_{zy})$ in the metric defined by Q_{yy}^{-1}.

Proof

The Lagrangian function of the inner minimization problem is given by

$$(3.14) \qquad L = (Y-\eta)' Q_{yy} (Y-\eta) - 2\kappa'(B_y'\eta + B_z'Z)$$

with κ the $(p \times 1)$ vector of Lagrange multipliers. Analogous to the proof for
the principal relations analysis in section 1, the optimal values
$\tilde{Y}$ and κ_o are obtained by setting the partial derivatives of L with respect to
η and κ equal to zero, yielding

$$(3.15) \qquad \kappa_o = -(\beta_y'Q_{yy}^{-1}B_y)^{-1}(B_y'Y+B_z'Z) = -(B_y'Q_{yy}^{-1}B_y)^{-1}B'X$$

and

$$(3.16) \qquad \tilde{Y} = Y - Q_{yy}^{-1} B_y (B_y' Q_{yy}^{-1} B_y)^{-1} B'X.$$

The matrix $(B_y' Q_{yy}^{-1} B_y)$ is non-singular (and therefore invertible) as we assumed $p < m$.

Using result (3.16), the outer minimization problem can be rewritten as

$$(3.17) \qquad \min_B E(X'B(B_y' Q_{yy}^{-1} B_y)^{-1} B'X).$$

This minimization problem can be rewritten in a more familiar form (which has also been observed in section 1 for the PR case) by using the Moore-Penrose inverse of semi-metric defining matrix $Q_o = \begin{bmatrix} Q_{yy} & 0 \\ \hline 0 & 0 \end{bmatrix}$, that is given by

$$(3.18) \qquad Q_o^+ = \begin{bmatrix} Q_{yy}^{-1} & 0 \\ \hline 0 & 0 \end{bmatrix} \qquad\qquad (k \times k).$$

Replacing $B_y' Q_{yy}^{-1} B_y$ by $B Q_o^+ B$, using trace operator properties and taking expectations, (3.17) can be rewritten as

$$(3.19) \qquad \min_B tr(B(B'Q_o^+ B)^{-1} B' \Sigma).$$

As the expression in (3.19) equals minimization problem (1.23) with Q^{-1} replaced by Q_o^+, the derivative of the objective function of this mathematical programming problem corresponds with the derivatives obtained in (1.24). Setting the derivatives equal to zero, the optimal value B_{se} of B is obtained as a solution of matrix equation

$$(3.20) \qquad (I_k - Q_o^+ B_{se}(B_{se}' Q_o^+ B_{se})^{-1} B_{se}') \Sigma B_{se}(B_{se}' Q_o^+ B_{se})^{-1} = 0.$$

Post-multiplication of (3.20) by $(B_{se}' Q_o^+ B_{se})$ and application of the partitionings of the matrices used in (3.20) yields

(3.21)
$$\left[\begin{array}{c|c} I_m - Q_{yy}^{-1} B_{se,y} (B'_{se,y} Q_{yy}^{-1} B_{se,y})^{-1} B'_{se,y} & -Q_{yy}^{-1} B_{se,y} (B'_{se,y} Q_{yy}^{-1} B_{se,y})^{-1} B'_{se,z} \\ \hline 0 & I_{k-m} \end{array}\right]$$

$$\times \left[\begin{array}{c} \Sigma_{yy} B_{se,y} + \Sigma'_{zy} B_{se,z} \\ \hline \Sigma_{zy} B_{se,y} + \Sigma_{zz} B_{se,z} \end{array}\right] = 0.$$

The lower part of (3.21) yields

(3.22) $\Sigma_{zy} B_{se,y} + \Sigma_{zz} B_{se,z} = 0$

and consequently the solution of $B_{se,z}$ is given as a linear function of $B_{se,y}$ by

(3.23) $B_{se,z} = -\Sigma_{zz}^{-1} \Sigma_{zy} B_{se,y}.$

Substitution of (3.20) in the upper part of (3.21) yields

(3.24) $(I_m - Q_{yy}^{-1} B_{se,y} (B'_{se,y} Q_{yy}^{-1} B_{se,y})^{-1} B'_{se,y})(\Sigma_{yy} - \Sigma'_{zy} \Sigma_{zz}^{-1} \Sigma_{zy}) B_{se,y} = 0.$

This relationship is fulfilled when the columns of $B_{se,y}$ equal p of the m characteristic vectors of $(\Sigma_{yy} - \Sigma'_{zy} \Sigma_{zz}^{-1} \Sigma_{zy})$ in the metric defined by Q_{yy}^{-1}. Hence $B_{se,y}$ is given as a solution of the system of equations

(3.25) $(\Sigma_{yy} - \Sigma'_{zy} \Sigma_{zz}^{-1} \Sigma_{zy}) B_{se,y} = Q_{yy}^{-1} B_{se,y} \Omega_p$

where Ω_p is a diagonal matrix with p of the m characteristic roots of $(\Sigma_{yy} - \Sigma'_{zy} \Sigma_{zz}^{-1} \Sigma_{zy})$ in the Q_{yy}^{-1}-metric on its diagonal. Again (like in the principal relations analysis) we do not yet know which characteristic vectors constitute matrix $B_{se,y}$. Substituting B_{se} in the objective function of (3.19) we obtain its value

(3.26) $\mathrm{tr}((B'_{se,y} Q_{yy}^{-1} B_{se,Y})^{-1} B_{se} \Sigma B_{se}) = \mathrm{tr}\, \Omega_p,$

which is at a minimum when we take the minimal characteristic roots as diagonal elements of Ω_p. ◊

The reader will notice the similarities between the SE analysis and the "basic" PR analysis dealt with in section 1. Not only is the SE analysis a special case of the PR analysis by assuming a specific metric defined by the matrix given in (3.6), but also the proof of this last theorem mainly follows the course of the proof on the PR matrix. Differences are introduced when matrix Q_o^+ is used (formula (3.20)) instead of having matrix Q^{-1}, like in matrix equation (1.24). Note that Q_o^+ can be seen as the inverse of the infinite metric matrix (3.6) if we are willing to accept the convention that $0 \times \infty = 1$. Combining matrix equations (3.23) and (3.25), the solution B_{se} is given by matrix equation

$$(3.27) \qquad \Sigma B_{se} = Q_o^+ B_{se} \Omega_{min,p}.$$

Because of the singularity of Q_o^+ we can no longer speak of "a metric defined by Q_o^+". Therefore the matrix B_{se} does not consist of columns that equal characteristic vectors, but an analytical solution for B_{se} had to be obtained by using the special form of Q_o^+, three of the four sub-matrices of which equal the null-matrix. Clearly because of the correspondence between the solutions B_{pr} and $B_{se,y}$ (they both consist of columns that are characteristic vectors of a matrix), the same remarks made for B_{pr} about orthogonality and normalization of its columns also hold for $B_{se,y}$. In particular the unit normalization (similar to (1.27)) of $B_{se,y}$ determines a unique solution such that

$$(3.28) \qquad B'_{se,y} Q_{yy}^{-1} B_{se,y} = I_p \quad \text{or} \quad B'_{se} Q_o^+ B_{se} = I_p.$$

Let us return to the under-identified case, when p=m. Where is the proof of the preceding theorem incorrect, when it is assumed that p=m? No difficulties arise until expression (3.21). This matrix equation simplifies to

$$(3.29) \qquad -(B_{se,z} B_{se,y}^{-1})'(\Sigma_{zy} B_{se,y} + \Sigma_{zz} B_{se,z}) = 0$$

$$(\Sigma_{zy} B_{se,y} + \Sigma_{zz} B_{se,z}) = 0.$$

This is a set of dependent relations in $B_{se,y}$ and $B_{se,z}$, such that the optimal solution B_{se} cannot uniquely be determined. Only the relation $B_{se,z} =$

$-\Sigma_{zz}^{-1}\Sigma_{zy}B_{se,y}$ follows from (3.29). This means that a description of the optimal linear subspace, that fits the distribution of X as close as possible, can be obtained, since we have $\tilde{Y} = -(B'_{se,y})^{-1}B'_{se,z}Z = \Sigma_{zy}\Sigma_{zz}^{-1}Z$. However, there are many relations in $\tilde{Y}$ and Z that describe this linear subspace. Note that we can distinguish two forms of identification. First the identification of the optimal subspace (geometrical identification) and second the identification of the set of relations that describe this subspace (algebraic identification). In order to let the set of relations, denoted by $B'_{se,y}\tilde{Y} + B'_{se,z}Z = 0$, be algebraically identified, some additional assumptions on the form of B_{se} are needed. From the preceding theorem it follows that the optimal value of B_y is (geometrically) given by p independent vectors of length p, that form a basis of the linear vector space $\mathrm{span}(\Sigma_{yy}-\Sigma'_{zy}\Sigma_{zz}^{-1}\Sigma_{zy})$. In order to identify this basis algebraically, $\frac{1}{2}p(p+1)$ additional pieces of information are required: p restrictions are needed to fix the norm of the basis-vectors and $\binom{p}{2} = \frac{1}{2}p(p-1)$ restrictions are needed to determine the angles between each couple of vectors in the basis. For instance a unique solution for B_{se} can be obtained by presupposing Q^{-1}-orthonormal columns in sub-matrix $B_{se,y}$. That is, it is assumed that $B'_{se,y}Q_{yy}^{-1}B_{se,y} = I_p$. Note that because of the symmetry in Q_{yy}^{-1} this restriction contains $\frac{1}{2}p(p+1)$ additional pieces of information. In this case the solution equals the Q^{-1}-orthonormal basis of $\mathrm{span}(\Sigma_{yy}-\Sigma'_{zy}\Sigma_{zz}^{-1}\Sigma_{zy})$, which is given by the p (normalized) characteristic vectors of $(\Sigma_{yy}-\Sigma'_{zy}\Sigma_{zz}^{-1}\Sigma_{zy})$ in the metric defined by Q_{yy}^{-1}.

The normalization of B_y in the Q_{yy}^{-1}-metric is an arbitrary choice to secure algebraic identification. Many other a priori assumptions on the form of $B_{se,y}$ are possible. For instance, in econometric literature it is common practice to set the diagonal elements of $B_{se,y}$ equal to -1. However, this yields only p restrictions, while $\frac{1}{2}p(p+1)$ restrictions are needed for algebraic identification. An example of p^2 restrictions on $B_{se,y}$ is given in section III.4, where it is assumed (even before distance minimization), that $B_y = -I_p$, yielding a set of seemingly unrelated regressions (SUR).

We have already emphasized the striking correspondence between the restriction $B'_y\eta + B'_z Z = 0$ in (3.8) and the set of equations, that is in econometric theory known as a simultaneous equations system

$$(3.30) \qquad B'_y Y_n + B'_z z_n + \varepsilon_n = 0 \qquad (n=1,\ldots,N).$$

Simultaneous equations are specified for a set of observations, where we also distinguish between endogenous variables Y_n and exogenous variables z_n. The endogenous variables have outcome values that are assumed to be determined through the joint interaction with the exogenous variables within the system. The exogenous variables are determined outside the system. These equation systems stem from the idea of modelling economic data by a set of mathematical statistical relations. A classical example is for instance given by a partial equilibrium setting for a single commodity, where price, quantity demanded and quantity supplied are determined simultaneously (see for instance Judge et al. (1982, p. 338)). In these econometric models it is assumed that the number of relations equals the number of endogenous variables. A naive approach to estimating the parameters of a system of simultaneous equations is that of ordinary least squares (OLS). This approach applies the least squares method to each equation separately. Disadvantage of this approach is that the distinction between explanatory endogenous and exogenous variables is ignored. It can be shown that because of the dependency between the endogenous variables and the stochastic disturbance terms, the parameters of equations are biased and inconsistently estimated by the OLS technique (see for instance Haavelmo (1943, 1947) and Dhrymes (1980, p. 174)). In order to get estimators of the equation parameters free of inconsistencies, one employs the reduced form of the model. In our notation this means that $p=m$ and the p reduced form relations are given by

$$(3.31) \qquad Y_n = -(B'_y)^{-1} B'_z z_n - (B'_y)^{-1} \varepsilon_n \qquad\qquad (n=1,\dots,N)$$

where B_y is assumed to be a $(p \times p)$ non-singular matrix. Consistent estimators are obtained in two stages. First, least squares is used to estimate the reduced form and to approximate the endogenous variables y_n by related variables that are uncorrelated with the disturbance terms. In the second stage least squares is applied to the structural form (3.30), where the endogenous variables have been replaced by their first-stage approximations. This two-stage least squares (2SLS) method has been independently developed by Theil (1953, 1958) and Basmann (1957, 1959, 1960). Basmann referred to this technique as a "generalized classical linear estimator". From the second stage residuals, also the covariance matrix of the disturbance terms can be estimated. A third stage for estimating the simultaneous equations parameters can then be formed by using this estimated covariance matrix in the generalized least squares (GLS) estimation of the parameters of the structural equations. This three-stage least squares (3SLS) approach was first studied by Zellner and Theil (1962) and its efficiency has, among others, been investigated by Sargan (1964), Rothenberg and Leenders (1964) and Madansky (1964).

Let us return to the SE parameters, that are obtained as the optimal values of a distance minimization problem. We may also distinguish in this approach two stages. First, the vector Y of endogenous variables is projected on an (m-p) dimensional linear subspace, yielding $\tilde{Y}$. As the projection is parallel to Z=0, the exogenous variables vector Z and the residual vector given by $(Y-\tilde{Y})$ are uncorrelated. That is

$$(3.32) \quad E(Z(Y-\tilde{Y})') = E(ZY'-ZZ'\Sigma_{zz}^{-1}\Sigma_{zy})B_y(B_y'Q_{yy}^{-1}B_y)^{-1}B_y'Q_{yy}^{-1} = 0.$$

Consequently Y is transformed such that the covariance structure with Z remains unchanged: $E(Z\tilde{Y}') = E(ZY')$. This projection does not have a counterpart in the classical econometric approach of simultaneous equations systems, as it is always assumed that the number of endogenous variables equals the number of equations. The second step is that the expected squared distance between Y and $\tilde{Y}$, with $\tilde{Y}$ such that $B_y'\tilde{Y} + B_z'Z = 0$, is minimized. This corresponds in the sample with applying an OLS estimation technique to the transformed endogenous variables and the exogenous variables. The two parameter matrices can be identified separately by chosing $\frac{1}{2}p(p+1)$ restrictions for the columns of B_y. Comparing the classical simultaneous equation system with the minimum distance approach one should keep in mind that, although some of the variables of X do not vary under the projection (which are therefore considered as the exogenous variables of the system), the complete vector X retains its stochastic character.

In order to obtain estimators of the parameter matrix B_{se}, derived as an implicit function of Σ in the preceding theorem, we assume to have knowledge of a consistent and asymptotically normally distributed estimator $\hat{\Sigma}$ of covariance matrix Σ. Corresponding to the partitioning of Σ, the partitioning of its estimator $\hat{\Sigma}$ is given by

$$(3.33) \quad \hat{\Sigma} = \left[\begin{array}{c|c} \hat{\Sigma}_{yy} & \hat{\Sigma}_{zy} \\ \hline \hat{\Sigma}_{zy} & \hat{\Sigma}_{zz} \end{array} \right] \qquad (k \times k)$$

with $\hat{\Sigma}_{yy}$ an $(m \times m)$ and $\hat{\Sigma}_{zz}$ a $((k-m) \times (k-m))$ matrix. The MME of B_{se} is obtained by replacing in the relation that determines B_{se} (see (3.13)) the covariance matrix Σ by its sample analogue $\hat{\Sigma}$. In section 1, that described the basic formulation of PR analysis for a non-singular Q, it was shown that the metric matrix Q can be factorized as $Q = Q^{\frac{1}{2}} Q^{\frac{1}{2}}$ and that the linear transformation of X by $Q^{\frac{1}{2}}$ yields a re-formulation of the distance minimizing problem in the Euclidean metric defined by I_k. In the same way an $(m \times m)$ matrix $Q_{yy}^{\frac{1}{2}}$, say, can be obtained such that

$$(3.34) \qquad Q_{yy}^{\frac{1}{2}} Q_{yy}^{\frac{1}{2}} = Q_{yy} \quad \text{and} \quad Q_{yy}^{\frac{1}{2}} Q_{yy}^{-1} Q_{yy}^{\frac{1}{2}} = I_m.$$

(Compare these properties with (1.11)). Consider the following transformation of vector X

$$(3.35) \qquad X^* = Q^{\frac{1}{2}} X \quad \text{with} \quad Q^{\frac{1}{2}} = \left[\begin{array}{c|c} Q_{yy}^{\frac{1}{2}} & 0 \\ \hline 0 & I_{k-m} \end{array} \right].$$

As a consequence the covariance matrix of X^* equals

$$(3.36) \qquad \Sigma^* = Q^{\frac{1}{2}} \Sigma Q^{\frac{1}{2}} = \left[\begin{array}{c|c} Q_{yy}^{\frac{1}{2}} \Sigma_{yy} Q_{yy}^{\frac{1}{2}} & Q_{yy}^{\frac{1}{2}} \Sigma_{zy}' \\ \hline \Sigma_{zy} Q_{yy}^{\frac{1}{2}} & \Sigma_{zz} \end{array} \right]$$

and the matrix, the characteristic vectors of which are needed, is given by

$$(3.37) \qquad Q_{yy}^{\frac{1}{2}} (\Sigma_{yy} - \Sigma_{zy}' \Sigma_{zz}^{-1} \Sigma_{zy}) Q_{yy}^{\frac{1}{2}}.$$

The distance between Y and η in the Q_{yy}-metric equals the distance between Y^* $(= Q_{yy}^{\frac{1}{2}} Y)$ and η^* $(= Q_{yy}^{\frac{1}{2}} \eta)$ in the I_m-metric. Defining $B_{se,y}^*$ and $\Omega_{min,p}^*$ to be the characteristic vectors and the diagonal matrix of characteristic roots respectively of matrix (3.37) in the I_m-metric, that is

$$(3.38) \qquad Q_{yy}^{\frac{1}{2}} (\Sigma_{yy} - \Sigma_{zy}' \Sigma_{zz}^{-1} \Sigma_{zy}) Q_{yy}^{\frac{1}{2}} B_{se,y}^* = B_{se,y}^* \Omega_{min,p}^*,$$

it follows that the optimal matrix $B_{se,y}$ can be obtained by

(3.39) $B_{se,y} = Q_{yy}^{\frac{1}{2}} B_{se,y}^*$, $B_{se,z} = B_{se,z}^*$ and $\Omega_{min,p} = \Omega_{min,p}^*$.

Consequently the transformed parameter matrix $B_{se,y}^*$ is obtained as a matrix of characteristic vectors in the simple Euclidean metric defined by I_m and the matrix $B_{se,y}$ of interest follows directly from (3.39). The facorization of Q_{yy} yields the following algorithm to compute the MME $\hat{B}_{se}$ of B_{se}.

1. compute $\hat{\Sigma}$ (asymptotically normal estimator of Σ);
2. deduce from Q_{yy} its characteristic roots and vectors;
3. compute $Q_{yy}^{\frac{1}{2}}$ as indicated in section 1;
4. deduce from $Q_{yy}^{\frac{1}{2}} (\hat{\Sigma}_{yy} - \hat{\Sigma}_{zy}' \hat{\Sigma}_{zz}^{-1} \hat{\Sigma}_{zy}) Q_{yy}^{\frac{1}{2}}$ its p minimal characteristic roots and the corresponding characteristic vectors, yielding $\hat{\Omega}_{min,p}$ and $\hat{B}_{se,y}^*$;
5. $\hat{B}_{se,y} := Q_{yy}^{\frac{1}{2}} \hat{B}_{se,y}^*$;
6. $\hat{B}_{se,z} := -\hat{\Sigma}_{zz}^{-1} \hat{\Sigma}_{zy} \hat{B}_{se,y}$.

Because of the asymptotic normality of $(vec\hat{\Sigma})$ it can, by means of the delta method, be shown that the distribution of $vec\hat{B}_{se}$ is asymptotically normal too, and the only difficulty is to deduce partial derivatives of $vec\hat{B}_{se}$ with respect to $(vec\Sigma)'$. As B_{se} consists of p columns that are all obtained in the same way as a function of Σ, it is sufficient to obtain the derivative of one column β_{se}, say, of B_{se}.

Theorem (the derivatives of the SE matrix)
Let ω and β_{se} be a scalar and a $(k \times 1)$ vector defined by the normal equations

(3.40) $\Sigma\beta_{se} = Q_o^+ \beta_{se} \omega$ and $\beta_{se}' Q_o^+ \beta_{se} = 1$

with Q_o^+ given in (3.18) and β_{se} unique. Then vector β_{se} may be considered as a differentiable function of Σ with

(3.41) $\nabla(\beta_{se};\Sigma) = \begin{bmatrix} \nabla(\beta_{se,y};\Sigma) \\ \hline \nabla(\beta_{se,z};\Sigma) \end{bmatrix}$

where

(3.42) $\nabla(\beta_{se,y};\Sigma) = -(\beta'_{se} \otimes [Q^{\frac{1}{2}}_{yy}((\Sigma^*_{yy}-\Sigma^*_{zy}\Sigma^{-1}_{zz}\Sigma^*_{zy})-\omega I_m)^+Q^{\frac{1}{2}}_{yy} \mid 0])$ $(m \times k^2)$

(3.43) $\nabla(\beta_{se,z};\Sigma) = -\Sigma^{-1}_{zz}\Sigma_{zy}\nabla(\beta_{se,y};\Sigma) - (\beta'_{se} \otimes [0 \mid \Sigma^{-1}_{zz}])$ $((k-m) \times k^2)$

with $\Sigma^*_{yy} = Q^{\frac{1}{2}}_{yy}\Sigma_{yy}Q^{\frac{1}{2}}_{yy}$ and $\Sigma^*_{zy} = \Sigma_{zy}Q^{\frac{1}{2}}_{yy}$. The null-matrices used in (3.42) and (3.43) are defined to be of dimensions $(m \times (k-m))$ and $((k-m) \times m)$ respectively.

The proof of this theorem mainly follows the procedure applied in section 1 in order to obtain the derivatives of β_{pr}. However, now the computations have to be adjusted, because of the typical form of metric matrix Q (see (3.6)).

<u>Proof</u>

From (3.40) it follows that

(3.44) $\beta'_{se}\Sigma\beta_{se} = \omega$

and substitution of (3.44) into the first equation of (3.40) yields

(3.45) $(I_k - Q^+_o\beta_{se}\beta'_{se})\Sigma\beta_{se} = 0.$

This brings us in a similar situation as defined in section 1 for the PR case (see (1.39)), with β_{pr} and Q^{-1} replaced by β_{se} and Q^+_o respectively. Analogous to the derivation of (1.45) it can be shown that

(3.46) $(\beta'_{se} \otimes (I_k - Q^+_o\beta_{se}\beta'_{se})) + ((\Sigma-\omega Q^+_o) - 2Q^+_o\beta_{se}\beta'_{se}\Sigma) \nabla(\beta_{se};\Sigma) = 0.$

Using the partitioning definitions for β_{se}, Σ and Q^+_o, and the fact that $\beta_{se,z} = -\Sigma^{-1}_{zz}\Sigma_{zy}\beta_{se,y}$, (3.46) can be rewritten as a combination of two matrix equations, viz.

(3.47) $(\beta'_{se} \otimes ([I_m \mid 0] - Q^{-1}_{yy}\beta_{se,y}\beta'_{se})) +$

$+ ((\Sigma_{yy}-\omega Q^{-1}_{yy}) - 2Q^{-1}_{yy}\beta_{se,y}(\beta'_{se,y}\Sigma_{yy}+\beta_{se,z}\Sigma_{zy})) \nabla(\beta_{se,y};\Sigma)+$

$+ \Sigma'_{zy}\nabla(\beta_{se,z};\Sigma) = 0$

and

(3.48) $(\beta'_{se} \otimes [0 \mid I_{k-m}]) + \Sigma_{zy} V(\beta_{se,y};\Sigma) + \Sigma_{zz} V(\beta_{se,z};\Sigma) = 0.$

Pre-multiplication of (3.48) by Σ_{zz}^{-1} yields the relationship between $V(\beta_{se,y};\Sigma)$ and $V(\beta_{se,z};\Sigma)$ that has been proposed in (3.43).

Substituting (3.43) in (3.47) and replacing $\beta_{se,y}$ by $-\Sigma_{zz}^{-1}\Sigma_{zy}\beta_{se,z}$, (3.46) can be rewritten as

(3.49) $(\beta'_{se} \otimes ([I_m \mid 0] - Q_{yy}^{-1}\beta_{se,y}\beta_{se}'))$

$+ ((\Sigma_{yy} - \Sigma_{zy}'\Sigma_{zz}^{-1}\Sigma_{zy}) - \omega Q_{yy}^{-1} - 2Q_{yy}^{-1}\beta_{se,y}\beta_{se,y}'(\Sigma_{yy} - \Sigma_{zy}'\Sigma_{zz}^{-1}\Sigma_{zy})) \, V(\beta_{se,y};\Sigma) = 0.$

Note the correspondence between (1.45) and (3.49). The formula in (1.45) represents an expression in Σ and its characteristic root and vector (λ and β_{pr}) in the metric defined by Q^{-1}. The very similar matrix equation is given in (3.49) for $(\Sigma_{yy} - \Sigma_{zy}'\Sigma_{zz}^{-1}\Sigma_{zy})$ with the characteristic root and vector (ω and $\beta_{se,y}$) in the metric defined by Q_{yy}^{-1}. The replacement of I_k by $[I_m \mid 0]$ emphasizes the fact that the first matrix equation is in k dimensions, while the latter equation is in m dimensions. From (3.49) it follows that (analogous to the determination of (1.37) from (1.45))

(3.50) $V(\beta_{se,y};\Sigma) = -(\beta'_{se} \otimes \left[Q_{yy}^{\frac{1}{2}}(Q_{yy}^{\frac{1}{2}}(\Sigma_{yy} - \Sigma_{zy}'\Sigma_{zz}^{-1}\Sigma_{zy})Q_{yy}^{\frac{1}{2}} - \omega I_m)^+ Q_{yy}^{\frac{1}{2}} \mid 0 \right]),$

from which (3.42) follows directly. ◇

The limiting distribution of $\text{vec}\hat{B}_{se}$ follows from application of the delta method with the results of the preceding theorems and the asymptotic normality of $\text{vec}\hat{\Sigma}$. This section ends by providing the form of the asymptotic covariance matrix in case of an "ideal sample", i.e., when the sample consists of N i.i.d. observations on X. Again (like in the PR analysis, see end of section 1) the results will be given for one column of matrix $\hat{B}_{se}$. The expressions for the vectorized matrix $\hat{B}_{se}$ follow directly from the results for one column.

Theorem ($\hat{\beta}_{se}$ for random i.i.d. observations)
Let $\{X_n : n=1,\ldots,N\}$ be a sample of N i.i.d. random observations on X and let

$\hat{\beta}_{se}$ be the MME of β_{se} which is defined in (3.40). Then

$$(3.51) \qquad \sqrt{N}(\hat{\beta}_{se} - \beta_{se}) \xrightarrow{D}$$

$$N\left(0, \left[\begin{array}{c|c} S^+ & 0 \\ \hline -\Sigma_{zz}^{-1}\Sigma_{zy}S^+ & \Sigma_{zz}^{-1} \end{array}\right] E(XX'\beta_{se}\beta_{se}'XX') \left[\begin{array}{c|c} S^+ & -S^+\Sigma_{zy}'\Sigma_{zz}^{-1} \\ \hline 0 & \Sigma_{zz}^{-1} \end{array}\right]\right),$$

with S^+ defined by

$$(3.52) \qquad S^+ = Q_{yy}^{\frac{1}{2}}(\Sigma_{yy}^* - \Sigma_{zy}^{*'}\Sigma_{zz}^{-1}\Sigma_{zy}^* - \omega I_m)^+ Q_{yy}^{\frac{1}{2}} \qquad\qquad (m \times m).$$

Proof

It is known from section II.2 that in case of a sample of i.i.d. observations, the limiting distribution of the sample covariance matrix $\hat{\Sigma}$ is given by

$$(3.53) \qquad \sqrt{N}(\text{vec}\hat{\Sigma} - \text{vec}\Sigma) \xrightarrow{D} N(0, V)$$

with

$$(3.54) \qquad V = E(\text{vec}(XX'))(\text{vec}(XX'))' - (\text{vec}\Sigma)(\text{vec}\Sigma)'.$$

The gradient of β_{se} is obtained by combining the results derived in the preceding theorem (see (3.42) and (3.43)), yielding

$$(3.55) \qquad \nabla(\beta_{se};\Sigma) = -(\beta_{se}' \otimes \left[\begin{array}{c|c} S^+ & 0 \\ \hline -\Sigma_{zz}^{-1}\Sigma_{zy}S^+ & \Sigma_{zz}^{-1} \end{array}\right]).$$

Application of the delta method yields asymptotic normality of $\sqrt{N}(\hat{\beta}_{se} - \beta_{se})$ with mean zero. The covariance matrix of the limiting distribution is obtained by pre- and post-multiplication of (3.54) by (3.55) and its transpose. Using (A.11) four times these matrix multiplications result in

$$(3.56) \qquad \left[\begin{array}{c|c} S^+ & 0 \\ \hline -\Sigma_{zz}^{-1}\Sigma_{zy}S^+ & \Sigma_{zz}^{-1} \end{array}\right] (E(XX'\beta_{se}\beta_{se}'XX') - \Sigma\beta_{se}\beta_{se}'\Sigma) \left[\begin{array}{c|c} S^+ & -S^+\Sigma_{zy}'\Sigma_{zz}^{-1} \\ \hline 0 & \Sigma_{zz}^{-1} \end{array}\right].$$

Using

$$(3.57) \qquad [S^+ \mathbin{\vdots} 0] \Sigma \beta_{se} = S^+ (\Sigma_{yy} - \Sigma'_{zy} \Sigma_{zz}^{-1} \Sigma_{zy}) \beta_{se,y} = S^+ Q_{yy}^{-1} \beta_{se,y} \omega = 0$$

and

$$(3.58) \qquad \left[0 \mathbin{\vdots} \Sigma_{zz}^{-1} \right] \Sigma \beta_{se} = \left[\Sigma_{zz}^{-1} \Sigma_{zy} \mathbin{\vdots} I_{k-m} \right] \beta_{se} = \Sigma_{zz}^{-1} \Sigma_{zy} \beta_{se,y} + \beta_{se,z} = 0$$

it can be shown that

$$(3.59) \qquad \begin{bmatrix} S^+ & \vdots & 0 \\ \hline -\Sigma_{zz}^{-1} \Sigma_{zy} S^+ & \vdots & \Sigma_{zz}^{-1} \end{bmatrix} \Sigma \beta_{se} \beta'_{se} \Sigma \begin{bmatrix} S^+ & \vdots & -S^+ \Sigma'_{zy} \Sigma_{zz}^{-1} \\ \hline 0 & \vdots & \Sigma_{zz}^{-1} \end{bmatrix} = 0.$$

Combination of (3.56) and (3.59) yields the covariance matrix as proposed in (3.51). ◊

As a consequence, the asymptotic covariance matrix of $\hat{\beta}_{se,y}$ is given by

$$(3.60) \qquad \frac{1}{N} S^+ E(YX' \beta_{se} \beta'_{se} XY') S^+ \qquad\qquad (m \times m)$$

and denoting (3.60) by $\text{var}(\hat{\beta}_{se,y})$, the asymptotic covariance matrix of $\hat{\beta}_{se,z}$ is given by

$$(3.61) \qquad \Sigma_{zz}^{-1} \Sigma_{zy} (\text{var}(\hat{\beta}_{se,y})) \Sigma'_{zy} \Sigma_{zz}^{-1} + \frac{1}{N} \Sigma_{zz}^{-1} E(ZX' \beta_{se} \beta'_{se} XZ) \Sigma_{zz}^{-1} \qquad ((k-m) \times (k-m)).$$

In this section we discussed the minimum distance problem, where some parts of vector X were assumed to be fixed (the exogenous component Z) and some parts of X were free to vary (the endogenous component Y) in order to obtain a projection vector $\tilde{X} = (\tilde{Y}', Z')'$ in the linear subspace $B'\tilde{X} = 0$. It has been noted that in case m=p (i.e. the number of endogenous variables equals the number of relations) the problem of obtaining an optimal matrix B_{se} appears to be under-identified (see (3.29)). Therefore, the intuitively justifiable restriction $B'_y Q_{yy}^{-1} B_y = I_p$ was introduced, in order to obtain unique results for B_{se}. In the next section we shall assume for the outer minimization problem (with respect to B) an even stronger assumption, viz. $B_y = -I_p$. Hence B_y is fixed and we only need to obtain the optimal value for B_z as a function of Σ. This special case will appear to be strongly related with a set of linear regression equations, which has by Zellner been called a set of seemingly unrelated regression equations (1962).

III.4 Seemingly unrelated regressions

A special case of a set of simultaneous equations is obtained when the
parameter sub-matrix B_y is set equal to $-I_p$. That is, the number of relations
(p) is assumed to equal the number of endogenous variables in vector Y (m) and
each endogenous variable appears only once per relation. Also here, like in SE
analysis, the sub-vector Z of X, that consists of exogenous variables, is
assumed to be fixed in the distance minimization analysis. The problem is to
obtain the optimal parameter sub-matrix B_z as a function of the covariance
matrix Σ by solving the mathematical programming problem defined in (3.8),
with the additional restriction $B_y = -I_p$, i.e.

$$(4.1) \qquad \min_{\substack{B_z}} E[\ \min_{\substack{\eta \\ \eta=\beta_z'Z}} \ (Y-\eta)'Q_{yy}(Y-\eta)].$$

This means that we are interested in the projection of X on a linear subspace
defined by $\{\xi | \xi=(\eta',\zeta)',\ \eta=B_z'\zeta\}$ with the projection direction parallel to the
linear subspace Z=0 (i.e. perpendicular to Y=0). Substitution of $B_z'Z$ for η in
$(Y-\eta)$ makes the inner minimization superfluous, as the auxiliary vector η
vanishes from the objective function. In other words: $\tilde{Y}$ (the optimal value of
η) equals $B_z'Z$. The outer minimization sustains and reads

$$(4.2) \qquad \min_{\substack{B_z}} E(Y-B_z'Z)'Q_{yy}(Y-B_z'Z).$$

From (4.2) it follows that the mean squared Q_{yy}-distance between the
endogenous vector Y and a linear combination of the components of the
exogenous vector Z is minimized. The restriction $\eta = B_z'Z$ can be seen as the
"best" linear approximation of the components of Y by a linear combination of
the random vector Z of "explanatory" variables. The analysis of deriving the
best fitting hyperplane subject to the aforementioned conditions is called, in
accordance to the statistical regression theory, "Seemingly Unrelated
Regression" (SUR) analysis. The idea of jointly considering a set of
regression equations has first been discussed by Zellner (1962) and some
properties of the efficiency of the SUR estimators have been shown by Zellner,
Huang (1962). Zellner showed that, in comparison with separate estimation of
the p linear relations, a gain in efficiency for the parameter estimators can
be obtained by jointly considering all the equations. For further reading on

SUR one is referred to Kakwani (1967), Revankar (1974), Schmidt (1977) and Binkley (1981).

The optimal value of B_z in (4.2) is obtained by setting the derivatives of the objective function with respect to B_z equal to zero and the solution will be indicated by $B_{sur,z}$. The value of B_y equals $-I_p$ by assumption, i.e.

(4.3) $B_{sur,y} = -I_p.$

Theorem (the SUR matrix)

Let the mathematical programming problem be defined by (4.2) with

  Y a $(p \times 1)$ random vector

  Z a $((k-p) \times 1)$ random vector

  B_z a $((k-p) \times p)$ parameter matrix

  Q_{yy} a $(p \times p)$ symmetric, positive definite matrix.

The optimal value $B_{sur,z}$ of B_z is given by

(4.4) $B_{sur,z} = \Sigma_{zz}^{-1} \Sigma_{zy}$

where $\Sigma = \begin{bmatrix} \Sigma_{yy} & \vdots & \Sigma'_{zy} \\ --- & \vdots & --- \\ \Sigma_{zy} & \vdots & \Sigma_{zz} \end{bmatrix}$ is the covariance matrix of $\begin{bmatrix} Y \\ Z \end{bmatrix}$.

Proof

Before taking derivatives, (4.2) is rewritten (using trace properties) as

(4.5) $\min_{B_z} \operatorname{tr}((\Sigma_{yy} - B'_z \Sigma_{zy} - \Sigma'_{zy} B_z + B'_z \Sigma_{zz} B_z) Q_{yy}).$

The derivatives of the four terms in (4.5) are observed separately. Clearly, as the first term does not contain a factor B_z,

(4.6) $\dfrac{\partial \operatorname{tr}(\Sigma_{yy} Q_{yy})}{\partial B_z} = 0$ $((k-p) \times p).$

According to (A.50) and (A.49) we have

(4.7) $\quad \dfrac{\partial\ \mathrm{tr}(B'_z \Sigma_{zy} Q_{yy})}{\partial B_z} = \Sigma_{zy} Q_{yy}$

and

(4.8) $\quad \dfrac{\partial\ \mathrm{tr}(\Sigma'_{zy} B_z Q_{yy})}{\partial B_z} = \Sigma_{zy} Q_{yy}$

respectively. Using (A.48) the derivative of the last term is given by

(4.9) $\quad \dfrac{\partial\ \mathrm{tr}(B'_z \Sigma_{zz} B_z Q_{yy})}{\partial B_z} =$

$$= ((\mathrm{vec} I_p)' \otimes I_{k-p})(I_p \otimes \dfrac{\partial(B'_z \Sigma_{zz} B_z)}{\partial B_z}\ (Q_{yy} \otimes I_p))(\mathrm{vec} I_p \otimes I_p).$$

Using the product rule for matrix derivatives (A.41) and some Kronecker product properties, the right-hand side of (4.9) is rewritten as

(4.10) $\quad ((\mathrm{vec} I_p)' \otimes I_{k-p})(I_p \otimes P_{k-p,p}(\Sigma_{zz} B_z Q_{yy} \otimes I_p))(\mathrm{vec} I_p \otimes I_p)$

$\qquad + ((\mathrm{vec} I_p)' \otimes I_{k-p})(I_p \otimes (\mathrm{vec} \Sigma_{zz} B_z)(\mathrm{vec} Q_{yy})'))(\mathrm{vec} I_p \otimes I_p).$

Using (A.21) to rewrite the first term of (4.10), and using (A.12) and (A.13) to rewrite the second term, this long expression is together with the left-hand side of (4.9) simplified as

(4.11) $\quad \dfrac{\partial\ \mathrm{tr}(B'_z \Sigma_{zz} B_z Q_{yy})}{\partial B_z} = 2\Sigma_{zz} B_z Q_{yy}.$

Setting the sum of the derivatives given in (4.6), (4.7), (4.8) and (4.11) equal to zero, we obtain the following normal equation, that defines the optimal value $B_{\mathrm{sur},z}$ of B_z

(4.12) $\quad -2\Sigma_{zy} Q_{yy} + 2\Sigma_{zz} B_{\mathrm{sur},z} Q_{yy} = 0,$

from which (4.4) follows directly. ◇

It is worth noting, that the optimal value of B is an explicit function of covariance matrix Σ, and that the distance matrix Q_{yy} does not appear in $B_{sur,z}$. This means that the matrix Q_{yy} does not influence the parameter matrix $B_{sur,z}$. This may be compared with Zellner's results in SUR analysis. Zellner developed the simultaneous estimator of various regression vectors to gain efficiency by jointly considering all the equations. However, it can be shown that when for every linear regression equation the vector of explanatory variables are identical, this efficiency gain vanishes, as in this case the SUR estimator equals the OLS estimator of the p separate relations (see for instance Judge et al. (1982, p. 321)). This is exactly the case in minimization problem (4.2): every component Y_i of Y (i=1,...,p) is approximated by the "best" linear combination of one and the same vector Z. The optimal $((k-p)\times1)$ vectors β_i (for i=1,...,p) together form the columns of $B_{sur,z}$. In practice it may be interesting to observe models with different explanatory vectors for the various components of Y. This case can be obtained by setting some of the elements of columns of B_z equal to zero. This situation will be discussed in section 5 of this chapter.

When p=1, and hence the vector Y degenerates into one random variable, we consider only one approximation of Y (given X). This case can be interpreted as the structural linear regression model, in which one random variable Y is "explained" by a vector of (k-1) random explanatory variables. The case of random explanatory variables in the linear regression model has already been discussed by Van Praag (1978, 1981). He referred to this technique as a "Model-free regression" approach, as the asymptotic distribution of the regression vector is obtained without any relation assumptions. Also White (1980) and Chamberlain (1982) paid attention to this problem. White studied the linear Taylor approximation of unknown regression functions with random variables and Chamberlain observed a distributed lag regression model in a panel data context.

A consistent estimator of $B_{sur,z}$ is obtained as proposed in chapter II, by means of the method of moments. Defining $\hat{\Sigma}$ to be a consistent and asymptotically normally distributed estimator of the covariance matrix, that is partitioned like Σ, the MME $\hat{B}_{sur,z}$ of $B_{sur,z}$ is given by

$$(4.13) \quad \hat{B}_{sur,z} = \hat{\Sigma}_{zz}^{-1}\hat{\Sigma}_{zy}.$$

According to the delta method $\text{vec}\hat{B}_{\text{sur},z}$ is asymptotically normally distributed and in order to compute its asymptotic covariance matrix, the partial derivatives of $\text{vec}B_{\text{sur},z}$ with respect to $(\text{vec}\Sigma)'$ are needed. The matrix of derivatives for $p=1$ (hence $B_{\text{sur},z}$ is a $((k-1)\times 1)$ vector) is already given by Van Praag (1978, 1981). In his papers he assumed $B_{\text{sur},z}$ to be a function of matrix Σ_{zz} and vector Σ_{zy}. Therefore he derived the partial derivatives of $B_{\text{sur},z}$ with respect to $(\text{vec}\Sigma_{zz})'$ and Σ'_{zy}. Corresponding results are given in the more general setting of $p \geq 1$ in the following theorem. In order to keep the notation consistent with the foregoing results, derivatives are obtained with respect to all covariances $(\text{vec}\Sigma)'$.

Theorem (the derivatives of the SUR matrix)

Let $B_{\text{sur},z}$ be defined as indicated in (4.4), then $B_{\text{sur},z}$ may be considered as a differentiable function of Σ with

$$(4.14) \qquad \nabla(B_{\text{sur},z};\Sigma) = -(B'_{\text{sur}} \otimes \left[0 \mid \Sigma_{zz}^{-1} \right]) \qquad\qquad ((k-p)p\times k^2)$$

with 0 denoting a $((k-p)\times p)$ null-matrix and

$$(4.15) \qquad B_{\text{sur}} = \begin{bmatrix} B_{\text{sur},y} \\ \hline B_{\text{sur},z} \end{bmatrix} = \begin{bmatrix} -I_p \\ \hline \Sigma_{zz}^{-1}\Sigma_{zy} \end{bmatrix}.$$

Proof

Before taking derivatives, the vector $\text{vec}B_{\text{sur},z} = \text{vec}(\Sigma_{zz}^{-1}\Sigma_{zy})$ will be rewritten, in such a way that only the result for matrix derivative $\dfrac{\partial\Sigma}{\partial(\text{vec}\Sigma)'}$ (which is given in (A.47)) is needed to obtain expression (4.14) as proposed in the theorem. For this reason (A.45) is used to rewrite $\nabla(B_{\text{sur},z};\Sigma)$ as

$$(4.16) \qquad \frac{\partial(\text{vec}(\Sigma_{zz}^{-1}\Sigma_{zy}))}{\partial(\text{vec}\Sigma)'} = (I_p \otimes \frac{\partial(\Sigma_{zz}^{-1}\Sigma_{zy})}{\partial(\text{vec}\Sigma)'})(\text{vec}I_p \otimes I_{k^2}).$$

The remaining matrix derivative to be assessed is split up into a sum of two matrix derivatives by means of the product rule (A.41) for matrix derivatives:

$$(4.17) \quad \frac{\partial(\Sigma_{zz}^{-1}\Sigma_{zy})}{\partial(\text{vec}\Sigma)'} = \frac{\partial\Sigma_{zz}^{-1}}{\partial(\text{vec}\Sigma)'} \, (\Sigma_{zy} \otimes I_{k^2}) + \Sigma_{zz}^{-1} \frac{\partial\Sigma_{zy}}{\partial(\text{vec}\Sigma)'}.$$

Using matrix derivative rule (A.52) for inverse matrices and rewriting Σ_{zz} as

$$(4.18) \quad \Sigma_{zz} = \left[0 \;\vdots\; I_{k-p}\right] \Sigma \left[\begin{array}{c} 0 \\ \hline I_{k-p} \end{array}\right]$$

and Σ_{zy} as

$$(4.19) \quad \Sigma_{zy} = \left[0 \;\vdots\; I_{k-p}\right] \Sigma \left[\begin{array}{c} I_p \\ \hline 0 \end{array}\right]$$

with the null-matrices of the appropriate dimensions, so that the matrix multiplications are meaningful, (4.17) may be rewritten as

$$(4.20) \quad \frac{\partial(\Sigma_{zz}^{-1}\Sigma_{zy})}{\partial(\text{vec}\Sigma)'} = -\left[0 \;\vdots\; \Sigma_{zz}^{-1}\right] \frac{\partial\Sigma}{\partial(\text{vec}\Sigma)'} \left(\left[\begin{array}{c} 0 \\ \hline B_{sur,z} \end{array}\right] \otimes I_{k^2}\right)$$

$$+ \left[0 \;\vdots\; \Sigma_{zz}^{-1}\right] \frac{\partial\Sigma}{\partial(\text{vec}\Sigma)'} \left(\left[\begin{array}{c} I_p \\ \hline 0 \end{array}\right] \otimes I_{k^2}\right)$$

$$= -\left[0 \;\vdots\; \Sigma_{zz}^{-1}\right] \frac{\partial\Sigma}{\partial(\text{vec}\Sigma)'} \, (B_{sur} \otimes I_{k^2}).$$

Replacing $\dfrac{\partial\Sigma}{\partial(\text{vec}\Sigma)'}$ by its outcome given in (A.47) and substituting (4.20) into (4.16) we obtain

$$(4.21) \quad \nabla(B_{sur,z};\Sigma) = -(I_p \otimes \left[0 \;\vdots\; \Sigma_{zz}^{-1}\right]((\text{vec}I_k)' \otimes I_k)(B_{sur} \otimes I_{k^2})) \times$$

$$(\text{vec}I_p \otimes I_{k^2}).$$

Using the elementary Kronecker properties (A.6) and (A.11), (4.21) can be rewritten as

$$(4.22) \quad \nabla(B_{sur,z};\Sigma) = -(I_p \otimes ((\text{vec}B_{sur})' \otimes \left[0 \;\vdots\; \Sigma_{zz}^{-1}\right]))(\text{vec}I_p \otimes I_{k^2})$$

and applying (A.6) and (A.13) this matrix expression can be reduced to
(4.14). ◇

As a result of the two preceding theorems we have

$$(4.23) \qquad \sqrt{N}(\text{vec}\hat{B}_{sur,z} - B_{sur,z}) \overset{D}{\rightarrow}$$

$$N(0, \ (B'_{sur} \otimes \begin{bmatrix} 0 & \vdots & \Sigma_{zz}^{-1} \end{bmatrix}) \ V \ (B_{sur} \otimes \begin{bmatrix} 0 \\ \hline \Sigma_{zz}^{-1} \end{bmatrix}))$$

with V the asymptotic covariance matrix of $\sqrt{N}(\text{vec}\hat{\Sigma} - \text{vec}\Sigma)$. The expression for
the covariance matrix at the right-hand side of (4.23) is considerably
simplified when it is assumed that $\hat{\Sigma}$ is obtained as the sample covariance
matrix of a random sample consisting of i.i.d. observations. In this case the
limiting distribution of $\text{vec}\hat{B}_{sur,z}$ is given by

<u>Theorem</u> ($\text{vec}\hat{B}_{sur,z}$ for random i.i.d. sample observations)
Let $\{X_n: n=1,\ldots,N\}$ be a sample of N i.i.d. random observations on X and let
$\hat{B}_{sur,z}$ be the MME of $B_{sur,z} = \Sigma_{zz}^{-1}\Sigma_{zy}$. Then

$$(4.24) \qquad \sqrt{N}(\text{vec}\hat{B}_{sur,z} - \text{vec}B_{sur,z}) \overset{D}{\rightarrow}$$

$$N(0, \ E[(Y-B'_{sur,z}Z)(Y-B'_{sur,z}Z)' \otimes \Sigma_{zz}^{-1}ZZ'\Sigma_{zz}^{-1}]).$$

<u>Proof</u>
All we need to show is the equivalence between the covariance matrices of the
limiting distributions given in (4.23) and (4.24) when V equals the covariance
matrix of vec(XX'). Using the partitioning of X and Σ, it is easy to see that
V can be written as

$$(4.25) \qquad V = E \begin{bmatrix} \text{vec}(XY') \\ \hline \text{vec}(XZ') \end{bmatrix} [(\text{vec}XY')' \vdots (\text{vec}XZ')'] - \begin{bmatrix} \text{vec}\Sigma_{xy} \\ \hline \text{vec}\Sigma_{xz} \end{bmatrix} [(\text{vec}\Sigma_{xy})' \vdots (\text{vec}\Sigma_{xz})']$$

where

$$(4.26) \qquad \Sigma_{xy} = \begin{bmatrix} \Sigma_{yy} \\ \hline \Sigma_{zy} \end{bmatrix} \quad \text{and} \quad \Sigma_{xz} = \begin{bmatrix} \Sigma'_{zy} \\ \hline \Sigma_{zz} \end{bmatrix}.$$

Pre- and postmultiplication of (4.25) by $(B'_{sur} \otimes \begin{bmatrix} 0 & \vdots & \Sigma_{zz}^{-1} \end{bmatrix})$ and its transpose yields for the asymptotic covariance matrix of $\sqrt{N}(\text{vec}\hat{B}_{sur,z} - \text{vec}B_{sur,z})$

$$(4.27) \qquad E(\text{vec}(\Sigma_{zz}^{-1}ZY'))(\text{vec}(\Sigma_{zz}^{-1}ZY'))' - (\text{vec}B_{sur,z})(\text{vec}B_{sur,z})'$$

$$- E(\text{vec}(\Sigma_{zz}^{-1}ZY'))(\text{vec}(\Sigma_{zz}^{-1}ZZ'B_{sur,z}))' + (\text{vec}B_{sur,z})(\text{vec}B_{sur,z})'$$

$$- E(\text{vec}(\Sigma_{zz}^{-1}ZZ'B_{sur,z}))(\text{vec}(\Sigma_{zz}^{-1}ZY'))' + (\text{vec}B_{sur,z})(\text{vec}B_{sur,z})'$$

$$+ E(\text{vec}(\Sigma_{zz}^{-1}ZZ'B_{sur,z}))(\text{vec}(\Sigma_{zz}^{-1}ZZ'B_{sur,z}))' - (\text{vec}B_{sur,z})(\text{vec}B_{sur,z})'.$$

Clearly the four terms $(\text{vec}B_{sur,z})(\text{vec}B_{sur,z})'$ cancel out as two of them appear with a positive sign and two of them with a negative sign. As the vec-operator is in the resulting terms applied to matrices that can be written as products of a column- and row vector, these vectorized matrices may be replaced by the Kronecker products of two column vectors. In effect, using (A.10), (4.27) can be rewritten as

$$(4.28) \qquad E(Y \otimes \Sigma_{zz}^{-1}Z)(Y' \otimes Z'\Sigma_{zz}^{-1}) - E(Y \otimes \Sigma_{zz}^{-1}Z)(Z'B_{sur,z} \otimes Z'\Sigma_{zz}^{-1})$$

$$- E(B'_{sur,z}Z \otimes \Sigma_{zz}^{-1}Z)(Y' \otimes Z'\Sigma_{zz}^{-1}) + E(B'_{sur,z}Z \otimes \Sigma_{zz}^{-1}Z)(Z'B_{sur,z} \otimes Z'\Sigma_{zz}^{-1}).$$

Using some elementary Kronecker product rules we obtain the following final expression for the asymptotic covariance matrix

$$(4.29) \qquad E[(Y-B'_{sur,z})(Y-B'_{sur,z})' \otimes \Sigma_{zz}^{-1}ZZ'\Sigma_{zz}^{-1}]. \quad \diamond$$

The matrix denoted in (4.29) is defined as the expectation of a Kronecker product of two random matrices, viz. $(Y-B'_{sur,z}Z)(Y-B'_{sur,z}Z)'$ and $\Sigma_{zz}^{-1}ZZ'\Sigma_{zz}^{-1}$. Realizing that (4.29) can be rewritten as the Kronecker product of the expectations of these two matrices when they are uncorrelated, it is not difficult to deduce the following proposition.

<u>Corollary</u> ($\text{vec}\hat{B}_{\text{sur},z}$ under the normality assumption)

Let $\{X_n : n=1,\ldots,N\}$ be a sample of N independent identically, normally $\mathbb{N}(0,\Sigma)$ distributed observations on X and let $\hat{B}_{\text{sur},z}$ be the MME of $B_{\text{sur},z}$. Then

$$(4.30) \qquad \sqrt{N}(\text{vec}\hat{B}_{\text{sur},z} - \text{vec}B_{\text{sur},z}) \overset{D}{\to} \mathbb{N}(0, ((\Sigma_{yy}-\Sigma'_{zy}\Sigma_{zz}^{-1}\Sigma_{zy}) \otimes \Sigma_{zz}^{-1})).$$

<u>Proof</u>

It is sufficient to show equivalence between (4.29) and the covariance matrix of the limiting distribution given in (4.30), for a normally distributed X. As

$$(4.31) \qquad E(Y-B'_{\text{sur},z}Z)Z' = \Sigma'_{zy}-\Sigma'_{zy}\Sigma_{zz}^{-1}\Sigma_{zz} = 0 \qquad\qquad (p\times(k-p))$$

the two vectors $(Y-B'_{\text{sur},z}Z)$ and Z are uncorrelated. From the normality assumption it follows that these vectors, and therefore also the products $(Y-B'_{\text{sur},z}Z)(Y-B'_{\text{sur},z}Z)'$ and ZZ', are mutually independent. Hence the covariance matrix of the limiting distribution is given by

$$(4.32) \qquad E(Y-B'_{\text{sur},z}Z)(Y-B'_{\text{sur},z}Z)' \otimes \Sigma_{zz}^{-1}E(ZZ')\Sigma_{zz}^{-1}$$

$$= (\Sigma_{yy}-\Sigma'_{zy}\Sigma_{zz}^{-1}\Sigma_{zy}) \otimes \Sigma_{zz}^{-1}. \quad \diamond$$

An estimator of (4.32) can be obtained by replacing the covariance matrix Σ by its sample analogue $\hat{\Sigma}$, and it is given by

$$(4.33) \qquad (\hat{\Sigma}_{yy}-\hat{\Sigma}'_{zy}\hat{\Sigma}_{zz}^{-1}\hat{\Sigma}_{zy}) \otimes \hat{\Sigma}_{zz}^{-1}.$$

It has already been mentioned that an important special case of the linear models is obtained when the parameter p in the SUR analysis is set equal to 1. In this case the estimator of $B_{\text{sur},z}$ corresponds with the OLS estimator of the classical linear regression model (which is a linear relation, the explanatory vector Z_n of which is assumed to be non-random), and the Kronecker product in (4.33) turns into a scalar multiplication as the first factor is a (p×p) matrix. Notice that the resulting expression corresponds with the estimated covariance matrix of the OLS estimator. Hence, not only the estimator

$\hat{B}_{sur,z}$ for p=1 equals the OLS estimator of the regression coefficients in the linear regression model, but under normality also their (asymptotic) covariance matrices are the same. Further evidence on the linear regression analysis, subject to various distributional assumptions for X can be found in Wesselman, Van Praag (1987).

III.5 Restricted seemingly unrelated regression

In the preceding section we observed the minimum distance parameters in a SUR context and we derived the limiting distributions of their MME. It appeared that the metric defining matrix Q_{yy} was not represented in the optimal solution $B_{sur,z}$, so that every column of the parameter matrix could be considered as the outcome of an isolated regression equation, in which each component of Y is approximated by a linear combination of some components of Z. This was caused by the fact that in each regression equation the vectors of explanatory variables are the same. In practice, often the situation is met with, that every component of Y is approximated by a different set of variables. This kind of problems can be described as a minimum distance problem by placing equality restrictions on the parameter matrix obtained by SUR analysis. To make this clear, we first consider a simple set of (seemingly unrelated) regressions, in which two endogenous variables are explained by three, respectively two exogenous variables, of wich only one appears in both relations. This example is denoted by

$$(5.1) \qquad Y_1 = \beta_{11}Z_1 + \beta_{21}Z_2 + \beta_{31}Z_3$$

$$Y_2 = \qquad\qquad\qquad + \beta_{32}Z_3 + \beta_{42}Z_4.$$

Let us rewrite (5.1) as

$$(5.2) \qquad Y = B_z'Z \quad \text{with}$$

$$Y' = (Y_1, Y_2), \quad Z' = (Z_1, Z_2, Z_3, Z_4), \quad B_z' = \begin{bmatrix} \beta_{11} & \beta_{21} & \beta_{31} & \beta_{41} \\ \beta_{12} & \beta_{22} & \beta_{32} & \beta_{42} \end{bmatrix}$$

and where B_z is restricted according to

$$(5.3) \qquad R'(\text{vec}B_z) = c \quad \text{with}$$

$$R' = \begin{bmatrix} 0 & 0 & 0 & 1 & 0 & 0 & 0 & 0 \\ 0 & 0 & 0 & 0 & 1 & 0 & 0 & 0 \\ 0 & 0 & 0 & 0 & 0 & 1 & 0 & 0 \end{bmatrix}, \quad c = \begin{bmatrix} 0 \\ 0 \\ 0 \end{bmatrix}.$$

Consequently a SUR model with different explanatory vectors can be obtained by adding restrictions to the minimum distance problem in the SUR analysis. As in restriction matrix R every column consists of only one non-zero element, each restriction is concerned with information on only one individual parameter. This type of restrictions can be extended to more general equality restrictions on linear combinations of parameters by defining more than one element per column of R unequal to zero. Furthermore, when R is a block-diagonal matrix, like in (5.3), then every restriction refers to a linear equality in one column of B_z. For instance when from prior information on B_z it is also known that in (5.1) β_{31} equals β_{32}, then this can be added to the model by extending R with column (0 0 1 0 0 0 -1 0) and c with an element equal to zero. In this section we shall derive the minimum distance solution for the SUR parameter matrix, where some prior information on the parameters is assumed in the form of exact linear relations between the components of B_z. The mathematical programming version of this problem is given by (see also (4.2))

$$(5.4) \quad \min_{B_z} E[(Y-B_z'Z)'Q_{yy}(Y-B_z'Z)]$$

$$\text{subject to } R'(vecB_z) = c$$

with R a $((k-p) \times r)$ matrix of rank r $(r \leq (k-p)p)$ and c an $(r \times 1)$ vector, both consisting of constants, that are known in advance. Naturally the number of independent restrictions (r) should not exceed the number of parameters $((k-p)p)$, as otherwise the restrictions are incompatible and the set of feasible solutions would be empty.

The use of prior information, like the linear equality restrictions as stated above, has already been discussed in the context of single linear regression equations by Durbin in 1953. Comprehensive treatments of restricted regression estimation techniques are presented in many econometric textbooks, like Judge et al. (1982, chapter 20) and Theil (1971, section 1.8). Attention to numerical problems in efficient estimation of constraint parameters is, for instance, given by Dent (1980) and Gallant and Gerig (1980). Extensions to sets of regression equations are treated by Theil and Goldberger (1961) in case of simultaneous equations and by Byron (1970) for SUR. The results obtained in the latter paper are comparable to the formulas obtained in the theorems of this section. One should keep in mind, however, that we work with structural linear models, while Byron discussed the functional analogue of it. Byron also applied his model to a (seemingly unrelated) system of demand equations, for which the parameters are estimated subject to the prior restrictions of classical demand theory.

The solution of the Restricted Seemingly Unrelated Regression (RSUR) problem, which is formulated in (5.4), is given by

<u>Theorem</u> (the RSUR matrix)

Let the mathematical problem be defined by (5.4) with

Y a $(p \times 1)$ random vector

Z a $((k-p) \times 1)$ random vector

B_z a $((k-p) \times p)$ parameter matrix

Q_{yy} a $(p \times p)$ symmetric, positive definite matrix

R a $((k-p)p \times r)$ matrix of fixed, known constants of rank r $(r \leq (k-p)p)$

c an $(r \times 1)$ vector of fixed, known constants.

The optimal value $B_{rsur,z}$ of B_z is given by

(5.5) $\text{vecB}_{rsur,z} = \text{vecB}_{sur,z} + (Q_{yy}^{-1} \otimes \Sigma_{zz}^{-1}) R \kappa_o$

with

(5.6) $\kappa_o = (R'(Q_{yy}^{-1} \otimes \Sigma_{zz}^{-1})R)^{-1}(c - R'(\text{vecB}_{sur,z}))$,

where $\Sigma = \begin{bmatrix} \Sigma_{yy} & \vdots & \Sigma_{zy}' \\ \cdots & \vdots & \cdots \\ \Sigma_{zy} & \vdots & \Sigma_{yy} \end{bmatrix}$ is the covariance matrix of $\begin{bmatrix} Y \\ Z \end{bmatrix}$ and $B_{sur,z} = \Sigma_{zz}^{-1}\Sigma_{zy}$.

<u>Proof</u>

The Lagrangian function of the minimization problem (5.4) is given by (see also (4.5))

(5.7) $L = \text{tr}((\Sigma_{yy} - B_z'\Sigma_{zy} - \Sigma_{zy}'B_z + B_z'\Sigma_{zz}B_z)Q_{yy}) - 2\kappa'(R'(\text{vecB}_z) - c)$

with κ the $(r \times 1)$ vector of Lagrange multipliers. The optimal values $B_{rsur,z}$ and κ_o of B_z and κ equal the stationary points of L and are determined by setting the partial derivatives of L with respect to B_z and κ equal to zero. The derivative of L with respect to κ yields the linear restriction for $vecB_{rsur,z}$, that is

(5.8) $\quad R'(vecB_{rsur,z}) = c.$

The derivative of the first term of L with respect to B_z has been obtained in the proof of the SUR matrix and is given by the sum of (4.6), (4.7), (4.8) and (4.11). A vectorized version of this sum is given by

(5.9) $\quad vec(-2\Sigma_{zy}Q_{yy} + 2\Sigma_{zz}B_zQ_{yy}).$

The vectorized version of the derivative of the second term of L is given by

(5.10) $\quad \dfrac{\partial(2\kappa'R'(vecB_z))}{\partial(vecB_z)} = 2R\kappa.$

Setting the sum of (5.9) and (5.10) equal to zero and combining the result with (5.8), we obtain the following set of equations that determine the optimal values

(5.11) $\quad vec((\Sigma_{zz}B_{rsur,z} - \Sigma_{zy})Q_{yy}) - R\kappa_o = 0$

(5.12) $\quad R'(vecB_{rsur,z}) = c.$

Rewriting (5.11) and (5.12) by means of (A.11) we get

(5.13) $\quad (Q_{yy} \otimes (\Sigma_{zz}B_{rsur,z} - \Sigma_{zy}))(vecI_p) - R\kappa_o = 0$

(5.14) $\quad R'(I_p \otimes B_{rsur,z})(vecI_p) = c.$

Pre-multiplication of (5.13) by $R'(Q_{yy}^{-1} \otimes \Sigma_{zz}^{-1})$ yields

(5.15) $\quad R'(I_p \otimes (B_{rsur,z} - B_{sur,z}))(vecI_p) - R'(Q_{yy}^{-1} \otimes \Sigma_{zz}^{-1})R\kappa_o = 0.$

Substituting (5.14) in (5.15) and again using (A.11) we obtain

(5.16) $(c-R'(vecB_{sur,z})) - R'(Q_{yy}^{-1} \otimes \Sigma_{zz}^{-1})R\kappa_o = 0$

from which (5.6) follows directly. Pre-multiplication of (5.13) by $(Q_{yy}^{-1} \otimes \Sigma_{zz}^{-1})$ and application of (A.11) yields (5.5).   ◇

Let us consider the resulting value for B_z more closely. The restricted optimal version of B_z is obtained by using both the information on the random vector X and the non-stochastic information given by the restriction. The resulting optimum differs from the unrestricted optimum $vecB_{sur,z}$ by a linear function of $(c-R'(vecB_{sur,z}))$. The correspondence with functional restricted regression analysis becomes clear when p is set equal to 1, so that $vecB_{rsur,z}$, $vecB_{sur,z}$ and Q_{yy} can be replaced by the $((k-1)\times1)$ vectors $\beta_{rsur,z}$, $\beta_{sur,z}$ and scalar 1 respectively, yielding

(5.17) $\beta_{rsur,z} = \beta_{sur,z} + \Sigma_{zz}^{-1}R\kappa_o$

and

(5.18) $\kappa_o = (R'\Sigma_{zz}^{-1}R)^{-1}(c-R'\beta_{sur,z}).$

Substitution of (5.18) in (5.17) yields

(5.19) $\beta_{rsur,z} = \beta_{sur,z} + \Sigma_{zz}^{-1}R(R'\Sigma_{zz}^{-1}R)^{-1}(c-R'\beta_{sur,z}).$

What is the geometrical interpretation of the RSUR analysis? Let us consider once more the solution for p=1, given in (5.19). In the unrestricted regression case the vector X was projected parallel to Z=0 on a linear subspace defined by $\{\xi|\xi=(\eta',\zeta')', \eta=\beta'_{sur,z}\zeta\}$. Hence the X vector is only transformed in the Y-dimension. When we place a homogeneous restriction on β_z (hence c=0), we may write (using (5.19))

(5.20) $\tilde{Y} = \beta'_{rsur,z}Z$

$= \beta'_{sur,z}(I_{k-1} - R(R'\Sigma_{zz}^{-1}R)^{-1}R'\Sigma_{zz}^{-1})Z.$

Or, in words, the restricted regression vector equals the unrestricted regression vector obtained from the regression of Y on a linearly transformed version of Z. From (5.20) it follows that this linear transformation is given by the orthogonal projection of Z on the linear subspace orthogonal to the subspace spanned by the columns of R in a metric defined by Σ_{zz}^{-1}. Hence, the homogeneously restricted linear regression is the transformation of a (k×1) vector X to a ((k-1)×1) linear subspace by the projection of the Z component orthogonal (in the Σ_{zz}^{-1}-metric) onto O(span(R)), yielding $\tilde{Z}$, combined with the projection of the Y component parallel to Z=0 (yielding $\tilde{Y}$), in such a way that $\tilde{Y} = \beta'_{sur,z} \tilde{Z}$. A non-homogeneous restriction (c≠0) causes, additional to the projection, a linear translation defined by $c'(R'\Sigma_{zz}^{-1}R)^{-1}R'\Sigma_{zz}^{-1}Z$. It follows that the restrictions in the SUR analysis bring about appealing generalizations of the SUR results. Not only in the formulas, but also in the geometrical interpretation. This is unfortunately not the case in PR analysis (as discussed in section 1), when minimum distance parameters are considered subject to linear equality restrictions. This case will not be discussed in this study. More on the relation between a priori (not necessarily linear) restriction specifications and resulting PR parameters can be found in Van Praag, Koster (1984).

In the beginning of this section, the restrictions on SUR were introduced in order to describe SUR models with different explanatory variables. As a result, the Q_{yy} matrix appeares in the optimal value of B_z (see (5.5)). However, this is not the case for every restriction matrix R. When the restrictions on $vecB_z$ are described by a set of the same restrictions for each column of B_z, then the set of RSUR can be considered as p isolated restricted regression equations. Implications of this special case for the formulas are given in the following proposition.

Corollary (p identical restrictions for the columns of B_z in RSUR analysis)
Let the $\bar{r}$ restrictions in the RSUR analysis per column of B_z be defined by

(5.21) $\bar{R}'\beta_{z,i} = \bar{c}_i$ for i=1,...,p

where $B_z = [\beta_{z,1},...,\beta_{z,p}]$ with $\beta_{z,i}$ a ((k-p)×1) vector,
   $\bar{R}$ is a ((k-p)×$\bar{r}$) matrix of constants of rank $\bar{r}$ ($\bar{r} \leq$ k-p),
   $\bar{c}_i$ is an ($\bar{r}$×1) vector of constants.

Then

(5.22) $\quad B_{rsur,z} = B_{sur,z} + \Sigma_{zz}^{-1}\bar{R}(\bar{R}'\Sigma_{zz}^{-1}\bar{R})^{-1}(\bar{C}-\bar{R}'B_{sur,z})$

with $\quad \bar{C} = [\bar{c}_1,\ldots,\bar{c}_p] \qquad\qquad (\bar{r}\times p).$

Proof

The preceding theorem can be applied, as we are in the RSUR case, with R, c and r defined by

(5.23) $\quad R = (I_p \otimes \bar{R})$, $c = vec\bar{C}$, $r = p\bar{r}.$

Substitution of (5.23) in (5.6) and application of (A.11) yields

(5.24) $\quad \kappa_o = ((I_p \otimes \bar{R}')(Q_{yy}^{-1} \otimes \Sigma_{zz}^{-1})(I_p \otimes \bar{R}))^{-1}(vec\bar{C} - (I_p \otimes \bar{R}')(vecB_{sur,z}))$

$\qquad = (Q_{yy}^{-1} \otimes \bar{R}'\Sigma_{zz}^{-1}\bar{R})^{-1}(vec(\bar{C}-\bar{R}'B_{sur,z}))$

$\qquad = vec\ ((\bar{R}'\Sigma_{zz}^{-1}\bar{R})^{-1}(\bar{C}-\bar{R}'B_{sur,z})Q_{yy}).$

Similarly, we obtain by substituting (5.23) in (5.5)

(5.25) $\quad vec\ B_{rsur,z} = vecB_{sur,z} + (Q_{yy}^{-1} \otimes \Sigma_{zz}^{-1})(I_p \otimes \bar{R})\kappa_o$

$\qquad = vecB_{sur,z} + (Q_{yy}^{-1} \otimes \Sigma_{zz}^{-1}\bar{R})\kappa_o.$

Substitution of (5.24) in (5.25) and application of (A.11) yields

(5.26) $\quad vecB_{rsur,z} = vecB_{sur,z} + vec(\Sigma_{zz}^{-1}\bar{R}(\bar{R}'\Sigma_{zz}^{-1}\bar{R})^{-1}(\bar{C}-\bar{R}'B_{sur,z}))$

from which (5.22) follows directly. $\diamond$

Having the parameter matrix of interest defined as a function of the covariance matrix Σ, an asymptotically normally distributed estimator can be derived straightforwardly. Again $\hat{\Sigma}$ is defined to be a consistent and asymptotically normally distributed estimator of Σ, and the asymptotic covariance matrix $\sqrt{N}(\text{vec}\hat{\Sigma} - \text{vec}\Sigma)$ is denoted by V. The MME $\hat{B}_{rsur,z}$ of $B_{rsur,z}$ is given by

$$(5.27) \quad \text{vec}\hat{B}_{rsur,z} = \text{vec}\hat{B}_{sur,z} +$$

$$(Q_{yy}^{-1} \otimes \hat{\Sigma}_{zz}^{-1})R(R'(Q_{yy}^{-1} \otimes \hat{\Sigma}_{zz}^{-1})R)^{-1}(c-R'(\text{vec}\hat{B}_{sur,z}))$$

$$\text{with } \hat{B}_{sur,z} = \hat{\Sigma}_{zz}^{-1}\hat{\Sigma}_{zy}.$$

Using the delta method, the limiting distribution of $\text{vec}\hat{B}_{rsur,z}$ is observed to be

$$(5.28) \quad \sqrt{N}(\text{vec}\hat{B}_{rsur,z} - \text{vec}B_{rsur,z}) \overset{D}{\to} N(0, [\nabla(B_{rsur,z};\Sigma)]V[\nabla(B_{rsur,z};\Sigma)]')$$

with the $((k-p)\times k^2)$ matrix $\nabla(B_{rsur,z};\Sigma)$ of partial derivatives of $\text{vec}B_{rsur,z}$ obtained in the following theorem.

<u>Theorem</u> (the derivative of $\text{vec}B_{rsur,z}$)

Let $B_{rsur,z}$ be the parameter matrix as denoted in (5.5) and (5.6), then $B_{rsur,z}$ may be considered as a differentiable function of Σ with

$$(5.29) \quad \nabla(B_{rsur,z};\Sigma) = [I_{(k-p)p} - (Q_{yy}^{-1} \otimes \Sigma_{zz}^{-1})R(R'(Q_{yy}^{-1} \otimes \Sigma_{zz}^{-1})R)^{-1}R']$$

$$\times (-B'_{rsur} \otimes \begin{bmatrix} 0 & \vdots & \Sigma_{zz}^{-1} \end{bmatrix})$$

with 0 denoting a $((k-p)\times p)$ null-matrix and

$$(5.30) \quad B_{rsur} = \begin{bmatrix} B_{rsur,y} \\ \hline B_{rsur,z} \end{bmatrix} = \begin{bmatrix} -I_p \\ \hline B_{rsur,z} \end{bmatrix}.$$

Note that the matrix of partial derivatives given in (5.29) is a product of two familiar matrices. The first factor is already known from the definition of $\text{vecB}_{\text{rsur},z}$ (see (5.5) and (5.6)) and the second factor resembles the matrix of partial derivatives in the unrestricted SUR (see (4.14)); the only difference is, that $B_{\text{sur},z}$ is replaced by $B_{\text{rsur},z}$.

Proof

The proof of this proposition consists of 2 parts. Firstly, the derivatives of κ_o, defined in (5.6), will be analysed and secondly, this result will be used in order to prove (5.29). Using the product rule for matrix derivatives (A.41), the matrix derivative of κ_o with respect to (vecΣ) is split up into two terms as

$$(5.31) \quad \frac{\partial \kappa_o}{\partial(\text{vec}\Sigma)'} = \frac{\partial(R'(Q_{yy}^{-1} \otimes \Sigma_{zz}^{-1})R)^{-1}}{\partial(\text{vec}\Sigma)'} [(c-R'(\text{vecB}_{\text{sur},z})) \otimes I_{k^2}]$$

$$+ (R'(Q_{yy}^{-1} \otimes \Sigma_{zz}^{-1})R)^{-1} \frac{\partial(c-R'(\text{vecB}_{\text{sur},z}))}{\partial(\text{vec}\Sigma)'}.$$

From the results of section 4 it follows that the second term of (5.31) can be denoted by

$$(5.32) \quad (R'(Q_{yy}^{-1} \otimes \Sigma_{zz}^{-1})R)^{-1}R'(B'_{\text{sur}} \otimes \left[0 \mid \Sigma_{zz}^{-1}\right]).$$

The first term of (5.31) is somewhat more laborious. Applying the derivative rule for inverse matrices (A.52), this first term is rewritten as

$$(5.33) \quad -(R'(Q_{yy}^{-1} \otimes \Sigma_{zz}^{-1})R)^{-1} \frac{\partial(R'(Q_{yy}^{-1} \otimes \Sigma_{zz}^{-1})R)}{\partial(\text{vec}\Sigma)'}$$

$$\times [(R'(Q_{yy}^{-1} \otimes \Sigma_{zz}^{-1})R)^{-1} \otimes I_{k^2}][(c-R'(\text{vecB}_{\text{sur},z})) \otimes I_{k^2}]$$

$$= -R'(Q_{yy}^{-1} \otimes \Sigma_{zz}^{-1})R)^{-1}R'(Q_{yy} \otimes \frac{\partial \Sigma_{zz}^{-1}}{\partial(vec\Sigma)'})(R\kappa_o \otimes I_{k^2}).$$

Similar to the deduction of (4.20) it can be shown that

$$(5.34) \qquad \frac{\partial \Sigma_{zz}^{-1}}{\partial(vec\Sigma)'} = -(vec \begin{bmatrix} 0 \\ \overline{\Sigma_{zz}^{-1}} \end{bmatrix})' \otimes \begin{bmatrix} 0 & \vdots & \Sigma_{zz}^{-1} \end{bmatrix}$$

with the null-matrices of the appropriate dimensions. Hence substituting (5.34) in (5.33) and applying some simple Kronecker product rules, the first term of (5.31) is shown to be

$$(5.35) \qquad (R'(Q_{yy}^{-1} \otimes \Sigma_{zz}^{-1})R)^{-1}R'((Q_{yy}^{-1} \otimes (vec \begin{bmatrix} 0 \\ \overline{\Sigma_{zz}^{-1}} \end{bmatrix})')(R\kappa_o \otimes \begin{bmatrix} 0 & \vdots & \Sigma_{zz}^{-1} \end{bmatrix})$$

$$= (R'(Q_{yy}^{-1} \otimes \Sigma_{zz}^{-1})R)^{-1}R'((B_{rsur,z} - B_{sur,z})' \otimes \begin{bmatrix} 0 & \vdots & \Sigma_{zz}^{-1} \end{bmatrix}).$$

Combination of the results obtained in (5.31), (5.32) and (5.35) yields

$$(5.36) \qquad \frac{\partial \kappa_o}{\partial(vec\Sigma)'} = -(R'(Q_{yy}^{-1} \otimes \Sigma_{zz}^{-1})R)^{-1}R'(-B_{rsur,z} \otimes \begin{bmatrix} 0 & \vdots & \Sigma_{zz}^{-1} \end{bmatrix}) \qquad (r \times k^2)$$

(with 0 denoting a ((k-p)×p) null-matrix), completing the first part of the proof.

The remaining part of the proof is not so hard to fulfill, as the matrix of partial derivatives in question is merely a combination of some preceding results. Using (5.5) and product rule (A.41) we have

$$(5.37) \qquad \nabla(B_{rsur,z};\Sigma) = \frac{\partial(vecB_{rsur,z})}{\partial(vec\Sigma)'} =$$

$$= \frac{\partial(vecB_{sur,z})}{\partial(vec\Sigma)'} + (Q_{yy}^{-1} \otimes \frac{\partial \Sigma_{zz}^{-1}}{\partial(vec\Sigma)'})(R\kappa_o \otimes I_{k^2}) + (Q_{yy}^{-1} \otimes \Sigma_{zz}^{-1})R \frac{\partial \kappa_o}{\partial(vec\Sigma)'}.$$

Replacing the three matrix derivatives at the right-hand side of (5.37) by their outcomes, that can be found in (4.14), (5.34) and (5.36) respectively, we obtain

$$(5.38) \quad \nabla(B_{rsur,z};\Sigma) = -(B'_{sur} \otimes \begin{bmatrix} 0 & \vdots & \Sigma_{zz}^{-1} \end{bmatrix}) - ((B'_{rsur}-B'_{sur}) \otimes \begin{bmatrix} 0 & \vdots & \Sigma_{zz}^{-1} \end{bmatrix})$$

$$+ (Q_{yy}^{-1} \otimes \Sigma_{zz}^{-1})R(R'(Q_{yy}^{-1} \otimes \Sigma_{zz}^{-1})R)^{-1}R'(B'_{rsur} \otimes \begin{bmatrix} 0 & \vdots & \Sigma_{zz}^{-1} \end{bmatrix}),$$

from which the result as proposed in (5.29) follows directly. ◊

From the results for $vecB_{rsur,z}$ (given in (5.5) and (5.6)) and its derivatives (given in (5.29)) it follows that the matrix

$$(5.39) \quad I_{(k-p)p} - (Q_{yy}^{-1} \otimes \Sigma_{zz}^{-1})R(R'(Q_{yy}^{-1} \otimes \Sigma_{zz}^{-1})R)^{-1}R'$$

plays an important role in RSUR analysis. It is not only the projection matrix that transforms $vecB_{sur,z}$ into $vecB_{rsur,z}$, but it also stands for the matrix that transforms $\nabla(B_{sur,z};\Sigma)$ into $\nabla(B_{rsur,z};\Sigma)$, after replacement of $B_{sur,z}$ by $B_{rsur,z}$ in the matrix of partial derivatives.

Let us again consider the special case of a single restricted regression. Setting p equal to 1, so that $vecB_{rsur,z}$, $vecB_{sur,z}$ and Q_{yy} can be replaced by $\beta_{rsur,z}$, $\beta_{sur,z}$ and scalar 1, the matrix of partial derivatives can be written as

$$(5.40) \quad \nabla(\beta_{rsur,z};\Sigma) = (I_{k-1} - \Sigma_{zz}^{-1}R(R'\Sigma_{zz}^{-1}R)^{-1}R')(\beta_{rsur,z} \otimes \begin{bmatrix} 0 & \vdots & \Sigma_{zz}^{-1} \end{bmatrix})$$

with 0 denoting the $((k-1)\times1)$ null-vector. Consequently, the covariance matrix of the limiting distribution of $\sqrt{N}(\hat{\beta}_{rsur,z} - \beta_{rsur,z})$ is obtained by pre- and post-multiplication of V by (5.40) (see also (5.28)). For the classical linear regression model, in which the vector Z of regressors is assumed to be non-random, the covariance matrix of the restricted regression vector is simply obtained by pre- en post-multiplication of the covariance matrix of the unrestricted regression vector by $(I_{k-1} - \hat{\Sigma}_{zz}^{-1}R(R'\hat{\Sigma}_{zz}^{-1}R)^{-1}R')$ and its transpose. See for instance Judge et al. (1982, pp. 552-556). On the other hand, it follows from (5.39), that in the structural case the resulting covariance matrix does nòt exactly equal the covariance matrix of $\sqrt{N}(\hat{\beta}_{sur,z} - \beta_{sur,z})$, pre- en post-multiplied by $(I_{k-1} - \Sigma_{zz}^{-1}R(R'\Sigma_{zz}^{-1}R)^{-1}R')$. The fact that we are now dealing with a structural linear equation (hence the explanatory vector Z is random in character, together with the variable Y to be explained), manifests itself by the substitution of $\beta_{rsur,z}$ for $\beta_{sur,z}$ in the original unrestricted regression covariance matrix.

Naturally, the matrix Q_{yy} is not in the resulting matrix of derivatives, if the restrictions are the same for every column of B_z. In this case the RSUR can be considered as p single restricted regressions and the results of the preceding theorem are simplified as

Corollary (the derivative of $vecB_{rsur,z}$ in case of p identical restrictions)
Let $B_{rsur,z}$ be defined as denoted in (5.22), then $B_{rsur,z}$ may be considered as a differentiable function of Σ with

$$(5.41) \qquad \nabla(B_{rsur,z};\Sigma) = (-B'_{rsur,z} \otimes \left[0 \mid \Sigma_{zz}^{-1} - \Sigma_{zz}^{-1}R(R'\Sigma_{zz}^{-1}R)^{-1}R'\Sigma_{zz}^{-1} \right]).$$

Proof
The results follow directly after substitution of the matrices defined in (5.21) in (5.29). ◇

This section ends with a theorem on the asymptotic normality of $vec\hat{B}_{rsur,z}$ in case of a sample consisting of N i.i.d. observations on X. Just like in the case of unrestricted SUR analysis, the covariance matrix of this limiting distribution can be specified explicitly, as the asymptotic covariance matrix V is known to equal the covariance matrix of vec(XX'). All we have to do, is to pre- and post-multiply this matrix by the matrix of partial derivatives of $vecB_{rsur,z}$ and its transpose. Before the results are formulated, some abbreviating notations are introduced, as otherwise the formulas would be to extensive. Let the matrix M be defined by

(5.42) $\quad M = (Q_{yy}^{-1} \otimes \Sigma_{zz}^{-1})R(R'(Q_{yy}^{-1} \otimes \Sigma_{zz}^{-1})R)^{-1} \qquad ((k-p)p \times r).$

Using (5.42) the most important results of the RSUR analysis, given in (5.5), (5.6) and (5.29), can be abbreviated as

(5.43) $\quad vecB_{rsur,z} = (I_{(k-p)p} - MR')vecB_{sur,z} + Mc$

and

(5.44) $\quad \nabla(B_{rsur,z};\Sigma) = (I_{(k-p)p} - MR')(B'_{rsur,z} \otimes \begin{bmatrix} 0 & \vdots & \Sigma_{zz}^{-1} \end{bmatrix}).$

An important property of M, that will be used in the proof of the next theorem is

(5.45) $\quad R'M = I_r.$

Theorem ($vec\hat{B}_{rsur,z}$ for random i.i.d. sample observations)
Let $\{X_n: n=1,\ldots,N\}$ be a sample of N i.i.d. random observations on X and let $\hat{B}_{rsur,z}$ be the MME of $B_{rsur,z}$ (defined in (5.43)). Then

(5.46) $\quad \sqrt{N}(vec\hat{B}_{rsur,z} - vecB_{rsur,z}) \xrightarrow{D}$

$$N(0, (I-MR')E[(Y-B'_{rsur,z}Z)(Y-B'_{rsur,z}Z)' \otimes \Sigma_{zz}^{-1}ZZ'\Sigma_{zz}^{-1}](I-MR')')$$

with M as proposed in (5.42).

From (4.24) it follows that this theorem constitutes a simple generalization of the results obtained in the unrestricted SUR regression.

Proof
The asymptotic normality of $vec\hat{B}_{rsur,z}$ has already been achieved in (5.28). We only have to compute the asymptotic covariance matrix, in case that V equals the covariance matrix of $vec(XX')$. Let the matrix V be partitioned as denoted in (4.25). Then the pre- and post-multiplication of V by the matrix of partial derivatives and its transpose, given in (5.44), yields

(5.47) $(I_{(k-p)p} - MR') \times$

$[E(\text{vec}(\Sigma_{zz}^{-1}ZY'))(\text{vec}(\Sigma_{zz}^{-1}ZY'))' - (\text{vecB}_{sur,z})(\text{vecB}_{sur,z})'$

$-E(\text{vec}(\Sigma_{zz}^{-1}ZY'))(\text{vec}(\Sigma_{zz}^{-1}ZZ'B_{rsur,z}))' + (\text{vecB}_{sur,z})(\text{vecB}_{rsur,z})'$

$-E(\text{vec}(\Sigma_{zz}^{-1}ZZ'B_{rsur,z}))(\text{vec}(\Sigma_{zz}^{-1}ZY'))' + (\text{vecB}_{rsur,z})(\text{vecB}_{sur,z})'$

$+E(\text{vec}(\Sigma_{zz}^{-1}ZZ'B_{rsur,z}))(\text{vec}(\Sigma_{zz}^{-1}ZZ'B_{rsur,z}))'-(\text{vecB}_{rsur,z})(\text{vecB}_{rsur,z})']$

$\times (I_{(k-p)p} - MR')'.$

Analogous to the method applied in section 4, to derive the asymptotic
covariance matrix of the SUR parameter matrix in case of an i.i.d. sample, it
can be shown that (5.47) equals

(5.48) $(I_{(k-p)p} - MR') \times$

$[E((Y-B'_{rsur,z}Z)(Y-B'_{rsur,z}Z)' \otimes \Sigma_{zz}^{-1}ZZ'\Sigma_{zz}^{-1})$

$- M(c-R'(\text{vecB}_{sur,z}))(c-R'(\text{vecB}_{sur,z}))'M']$

$\times (I_{(k-p)p} - MR')'.$

From property (5.45) it follows that the second term of the middle factor of
(5.48) vanishes when it is multiplied by $(I_{(k-p)p} - MR')$. The remaining matrix
product equals the covariance matrix of the limiting distribution proposed in
(5.46). ◊

Let us, in short, review the connections between the various sections of this
chapter. In section 1 the most general formulation of a minimum distance
problem in an Euclidean metric was studied. By defining the metric matrix Q of
a specific form, the principal relations take the form of a simultaneous
equations system with endogenous and exogenous variables. This case has been
studied in section 3. In section 4 the parameter matrix corresponding with the

endogenous variables was defined to equal the identity matrix, yielding a set
of (seemingly unrelated) regressions. By considering only one linear equation,
the SUR analysis degenerates into a single linear structural regression.
Finally, in this section we studied the SUR model subject to some additional
linear restrictions for the parameter matrix corresponding with the exogenous
variables. Consequently, in each section a new statistical technique is
obtained as a special case of the aforementioned method, by placing
supplementary restrictions on the metric matrix Q or the parameter matrix B.
The last section of this chapter also proceeds from the most general
formulation of the PR problem. It will be observed in which way the
statistical method, that is known under the name Canonical Correlation
analysis, is related with the minimum distance approach.

III.6 Canonical correlation analysis

In Canonical Correlation (CC) analysis two sets of random variables with a joint distribution are considered, and the correlations between the variables of one set and those of the other set are studied. The objective is to find a linear combination of the first set of variables, that has maximum correlation with a linear combination of the second set, subject to certain conditions. This theory was developed by Hotelling (1935, 1936). Formally, these sets of variables and their linear combinations can be denoted by Y, Z, and $B_y'Y$ and $B_z'Z$ respectively, with their dimensions in conformity with the definitions in the preceding sections. Again the random vectors Y and Z are together denoted by a (k×1) vector $X = (Y',Z')'$. However, it should be noted that this partitioning no longer represents exogeneity of the Z component. The partitioning is merely introduced to indicate that there are two sets of variables under consideration. The number of relations (in Y and Z) are indicated by p with $p \leq \min(m,k-m)$, where m and k-m are the number of variables in Y and Z respectively. The additional conditions that have to be satisfied by $B_y'Y$ and $B_z'Z$ are that the elements of $B_y'Y$ are pairwise uncorrelated, similarly, that the elements of $B_z'Z$ are pairwise uncorrelated and that the elements of the two linear compounds have unit variance. Formally, the problem is the following. Let the columns of B_y and B_z be defined by

(6.1) $\quad B_y = [\beta_{y,1},\dots,\beta_{y,p}]$ and $B_z = [\beta_{z,1},\dots,\beta_{z,p}]$.

Then the optimal values of B_y and B_z are obtained as the solutions of p mathematical programming problems, given for i=1,...,p by

(6.2) $\quad \max_{\beta_{y,i},\beta_{z,i}} \beta_{y,i}' \Sigma_{zy}' \beta_{z,i}$

$$\text{sub } \beta_{y,i}' \Sigma_{yy} \beta_{y,j} = \delta_{ij}$$

$$\beta_{z,i}' \Sigma_{zz} \beta_{z,j} = \delta_{ij} \qquad\qquad \text{for } j=1,\dots,i.$$

with δ_{ij} the Kronecker delta. From (6.2) it follows that the object of the ith maximization problem is to find the linear combinations of Y and Z with maximum correlation, subject to the $2(i-1)$ restrictions that $\beta'_{y,i}Y$ and $\beta'_{z,i}Z$ are uncorrelated with the previously obtained linear combinations (defined by the restrictions for $j=1,\ldots,i-1$), and that they both have unit variance (defined by the restrictions for $j=i$). The latter restrictions are necessary, since the optimal solution for multiples of $B'_y Y$ and $B'_z Z$ is the same as the optimal solution for $B'_y Y$ and $B'_z Z$ themselves. The unit variance normalization is arbitrary and is only made in order to obtain unique optimal solutions. In this study the problem, as formulated in (6.2), will not be worked out and for thourough treatment and detailed accounts on this problem one is referred to Anderson (1958, chapter 12), Van de Geer (1971, chapter 14), Morrison (1978, section 7.6) and Muirhead (1982, section 11.3). An extension of CC analysis to more than two sets of variables can for instance be found in Kettenring (1971). In what follows, CC will be formulated as a special case of PR analysis. In this way we are able to compare two seemingly completely different techniques like SE and CC with one another. The question is: how can CC analysis be formulated as a minimum distance problem? We shall start with a reformulation of (6.2) in the context of PR analysis, based on intuitively justifiable arguments. It will appear that the solution of the reformulated problem corresponds with the optimal values of the columns of B_y and B_z in the CC problem as stated in (6.2). Note that the two optimization problems are not claimed to be the same. They merely produce the same optimal value for the $(k\times p)$ parameter matrix $B = [B'_y \mathrel{\vdots} B'_z]'$.

In order to find similarities between the PR problem, given as a composite mathematical programming problem in (1.9), and the CC problem, defined in (6.2), the PR problem will be reconsidered with a specific metric defining matrix Q. Denoting this special choice by Q_{cc}, the metric in the CC version of the minimum distance problem is defined by

$$(6.3) \qquad Q_{cc} = \begin{bmatrix} \Sigma_{yy}^{-1} & \vdots & 0 \\ \cdots & \vdots & \cdots \\ 0 & \vdots & \Sigma_{zz}^{-1} \end{bmatrix}.$$

A complete formulation of the PR problem with the metric defined by Q_{cc} is now given by

$$(6.4) \quad \min_{\substack{B}} E[\min_{\substack{\xi \\ B'\xi=0}} (X-\xi)' \begin{bmatrix} \Sigma_{yy}^{-1} & 0 \\ \hline 0 & \Sigma_{zz}^{-1} \end{bmatrix} (X-\xi)].$$

The optimal value of the inner minimization problem does not change because of this choice of Q. Therefore, the outer minimization problem is similar to the minimization problem given in (1.23), which has been obtained by replacing the objective function of the inner minimization problem by its minimal value. Substituting Q_{cc} for Q, the resulting minimization problem is given by

$$(6.5) \quad \min_{\substack{B}} tr((B_y'\Sigma_{yy}B_y + B_z'\Sigma_{zz}B_z)^{-1}B'\Sigma B).$$

Before deriving the solution of this problem, let us consider the normalization of B in the Q^{-1}-metric. Recall that the normalization of B has been applied to its optimal value in order to guarantee unique identification. Here the normalization will be made prior to solving the minimization and is defined by

$$(6.6) \quad B_y'\Sigma_{yy}B_y = B_z'\Sigma_{zz}B_z = I_p,$$

yielding the familiar unit normalization with factor 2, that is

$$(6.7) \quad B'Q_{cc}^{-1}B = 2I_p.$$

Combining (6.5) with normalization (6.6) we obtain

$$(6.8) \quad \min_{\substack{B_y,B_z}} tr(I_p + B_y'\Sigma_{zy}'B_z)$$

$$\text{sub } B_y'\Sigma_{yy}B_y = B_z'\Sigma_{zz}B_z = I_p.$$

Introducing a minus sign in connection with B_z (hence the restriction $\beta_y'\eta + B_z'\zeta = 0$ for the inner minimization problem denoted in (6.4) is changed into $B_y'\eta = B_z'\zeta$), (6.8) may be rewritten as

(6.9) $\quad \underset{B_y, B_z}{\max} \quad tr(B_y' \Sigma_{zy}' B_z)$

$\quad\quad$ sub $B_y' \Sigma_{yy} B_y = B_z' \Sigma_{zz} B_z = I_p.$

Note the correspondence and difference between (6.2) and (6.9). Both
expressions describe the maximization of the correlations between
$B_y'Y$ and $B_z'Z$, however, in the first problem the correlations are minimized
successively, while in the latter problem the correlations are minimized
simultaneously. The conclusion is that the PR problem in the metric defined by
Q_{cc}, as denoted in (6.3) with a minus sign in connection with B_z, yields a
minimization problem that is closely related to the CC analysis. Solutions of
(6.4) are given in the following theorem.

Theorem (the CC matrix)
Let the mathematical programming problem be defined by (6.4), with

$\quad$ X = (Y',Z') $\quad$ a (k×1) random vector

$\quad$ Y,Z $\quad\quad\quad$ an (m×1) and a ((k-m)×1) random vector

$\quad$ ξ $\quad\quad\quad\quad$ a (k×1) auxiliary vector

$\quad$ B $\quad\quad\quad\quad$ a (k×p) parameter matrix of rank $p \le min(m, k-m)$,

and $\Sigma = \begin{bmatrix} \Sigma_{yy} & \vdots & \Sigma_{zy}' \\ \cdots & \vdots & \cdots \\ \Sigma_{zy} & \vdots & \Sigma_{zz} \end{bmatrix}$ the covariance matrix of X.

Then the optimal value $\tilde{X}$ of ξ is given by

(6.10) $\quad \tilde{X} = \begin{bmatrix} \tilde{Y} \\ \tilde{Z} \end{bmatrix}$ with

$\quad\quad \tilde{Y} = Y - \Sigma_{yy} B_y (B_y' \Sigma_{yy} B_y + B_z' \Sigma_{zz} B_z)^{-1} B'X$

$\quad\quad \tilde{Z} = Z - \Sigma_{zz} B_z (B_y' \Sigma_{yy} B_y + B_z' \Sigma_{zz} B_z)^{-1} B'X$

and the optimal value B_{cc} of B is defined by

$$(6.11) \qquad B_{cc} = \left[\begin{array}{c} B_{cc,y} \\ \hline B_{cc,z} \end{array}\right] \text{ with}$$

$$\Sigma'_{zy} \Sigma^{-1}_{zz} \Sigma_{zy} B_{cc,y} = \Sigma_{yy} B_{cc,y} \Phi^2_{max,p}$$

$$\Sigma_{zy} \Sigma^{-1}_{yy} \Sigma'_{zy} B_{cc,z} = \Sigma_{zz} B_{cc,z} \Phi^2_{max,p}$$

where $\Phi^2_{max,p}$ represents the $(p \times p)$ diagonal matrix with on its diagonal the p largest characteristic roots of $\Sigma'_{zy} \Sigma^{-1}_{zz} \Sigma_{zy}$ in the metric defined by Σ_{yy}, that equal the p largest characteristic roots of $\Sigma_{zy} \Sigma^{-1}_{yy} \Sigma'_{zy}$ in the metric defined by Σ_{zz}.

Proof

Comparing (6.4) with (1.9), it is to be noticed that we have a mathematical programming problem, that has already been solved in section 1. Only now the problem is stated in a metric defined by Q_{cc}, given in (6.3). Hence the solutions of (6.4) are obtained by rewriting (1.15) and (1.16) (the solutions of (1.9)), with Q replaced by Q_{cc}. From substitution of (6.3) in (1.15) directly follows (6.10). The substitution of (6.3) in (1.16) yields

$$(6.12) \qquad \Sigma_{yy} B_{cc,y} + \Sigma'_{zy} B_{cc,z} = \Sigma_{yy} B_{cc,y} \Lambda_{cc,min,p}$$

$$(6.13) \qquad \Sigma_{zy} B_{cc,y} + \Sigma_{zz} B_{cc,z} = \Sigma_{zz} B_{cc,z} \Lambda_{cc,min,p}$$

with $\Lambda_{cc,min,p}$ the $(p \times p)$ diagonal matrix with on its diagonal the p minimal characteristic roots of Σ in the metric defined by Q_{cc}. Post-multiplication of (6.12) by $(I_p - \Lambda_{cc,min,p})$, pre-multiplication of (6.13) by $\Sigma^{-1}_{zy} \Sigma^{-1}_{zz}$ and subtraction of the resulting expressions yields

$$(6.14) \qquad \Sigma'_{zy} \Sigma^{-1}_{zz} \Sigma_{zy} B_{cc,y} = \Sigma_{yy} B_{cc,y} (I_p - \Lambda_{cc,min,p})^2.$$

Denoting $(I_p - \Lambda_{cc,min,p})$ by $\Phi_{max,p}$, the first relation of (6.11) follows directly.

The second expression of (6.11) can be proven in a similar way. ◊

From (6.11) it follows that the solution for B_y is given by the characteristic vectors of $\Sigma'_{zy}\Sigma^{-1}_{zz}\Sigma_{zy}$ in a metric defined by Σ_{yy}, that correspond with the p largest characteristic roots. As $\Sigma'_{zy}\Sigma^{-1}_{zz}\Sigma_{zy}$ is of rank $\min(m,k-m)$ it is always possible to find p non-zero characteristic roots (recall: $p \leq \min(m,k-m)$). Analogous remarks can be made for $B_{cc,z}$. In order to let the characteristic vectors be uniquely determined, $B_{cc,y}$ and $B_{cc,z}$ are normalized as proposed in (6.6), so that

(6.15) $B'_{cc,y}\Sigma_{yy}B_{cc,y} = I_p$ and $B'_{cc,z}\Sigma_{zz}B_{cc,z} = I_p.$

Those, who are familiar with CC analysis, may have recognized the solution denoted in (6.11). It equals the optimal value obtained after solving maximization problem (6.2), and from (6.12) it follows that the (maximum) covariances between $B'_{cc,y}Y$ and $-B'_{cc,z}Z$ are given by

(6.16) $\text{cov}(B'_{cc,y}Y, -B'_{cc,z}Z) = \Phi_{max,p},$

which also corresponds with the results, known from classical CC analysis. The minus sign is used in connection with $B_{cc,z}$, as it was introduced to rewrite $B'_y\tilde{Y} = B'_z\tilde{Z}$ as $B'\tilde{X} = 0$. The linear compounds $B'_{cc,y}Y$ and $B'_{cc,z}Z$ are in the statistical literature known as the canonical variables and the diagonal elements of $\Phi_{max,p}$ are called the canonical correlations between Y and Z. We may conclude that an intuitively introduced minimum distance problem yields the same optimal values as the maximization problem denoted in (6.2), known from classical CC analysis.

From the minimum distance approach of CC analysis follows its relationship with simultaneous equations systems, as they both are special forms of determining principal relations in X. These correspondences between SE and CC do not yet follow from the resulting expressions for B_{se} and B_{cc}. Let us therefore consider the matrix B_{cc} more closely. B_{cc} was obtained as the solution of matrix equations (6.12) and (6.13). Using $\Phi_{max,p} = I_p - \Lambda_{cc,min,p}$, these relations are rewritten as

(6.17) $\Sigma_{yy}B_{cc,y}\Phi_{max,p} + \Sigma'_{zy}B_{cc,z} = 0$

(6.18) $\Sigma_{zy}B_{cc,y} + \Sigma_{zz}B_{cc,z}\Phi_{max,p} = 0.$

Pre-multiplication of (6.17) by $\Sigma_{zy}\Sigma_{yy}^{-1}$ and subtraction of (6.18) from the resulting expression yields

(6.19) $\Sigma_{zy}B_{cc,y}(\Phi_{max,p} - I_p) + \Sigma_{zy}\Sigma_{yy}^{-1}\Sigma_{zy}'B_{cc,z} - \Sigma_{zz}B_{cc,z}\Phi_{max,p} = 0.$

Replacing the middle term of (6.19) by $\Sigma_{zz}B_{cc,z}\Phi_{max,p}^{2}$ (see (6.11)), it follows that

(6.20) $\Sigma_{zy}B_{cc,y} + \Sigma_{zz}B_{cc,z}\Phi_{max,p} = 0,$

which may be rewritten as

(6.21) $B_{cc,z} = -\Sigma_{zz}^{-1}\Sigma_{zy}B_{cc,y}\Phi_{max,p}.$

Hence $B_{cc,z}$ can be defined as a linear function of $B_{cc,y}$. A similar relation has been obtained for the SE parameter matrix B_{se}, viz. (see (3.23)) $B_{se,z} = -\Sigma_{zz}^{-1}\Sigma_{zy}B_{se,y}$. These two relationships differ only a factor $\Phi_{max,p}$ (the diagonal matrix of canonical correlations), which is caused by the fact that in CC analysis X is not only projected in the Y dimensions, but also in the Z dimensions. Hooper (1959) derived some corresponding results. He accomplished a relation between classical canonical correlation coefficients and the estimators of simultaneous equations parameters in order to contribute towards the understanding of econometric analysis of simultaneous equations. Corresponding relationships have been observed by Yohai en Garcia Ben (1980) and Muller (1981) between SUR- and CC analysis. Some similarities between CC analysis and various other multivariate techniques, like Factor analysis, have been discussed by McKeon (1965). (In chapter IV of this study, Factor analysis will be presented in the context of minimum distance parameters).

Using the method of moments, a consistent and asymptotically normally distributed estimator $\hat{B}_{cc}$ of B_{cc} is defined by

(6.22) $\hat{\Sigma}_{yy}^{-1}\hat{\Sigma}_{zy}'\hat{\Sigma}_{zz}^{-1}\hat{\Sigma}_{zy}\hat{B}_{cc,y} = \hat{B}_{cc,y}\hat{\Phi}_{max,p}^{2}$

(6.23) $\hat{\Sigma}_{zz}^{-1}\hat{\Sigma}_{zy}\hat{\Sigma}_{yy}^{-1}\hat{\Sigma}_{zy}'\hat{B}_{cc,z} = \hat{B}_{cc,z}\hat{\Phi}_{max,p}^{2},$

where $\hat{\Sigma}$ denotes an asymptotically normally distributed estimator of Σ and $\hat{\Phi}_{max,p}$ is a diagonal matrix with on its diagonal the p maximal characteristic roots of $\hat{\Sigma}_{yy}^{-1}\hat{\Sigma}_{zy}'\hat{\Sigma}_{zz}^{-1}\hat{\Sigma}_{zy}$ (or $\hat{\Sigma}_{zz}^{-1}\hat{\Sigma}_{zy}\hat{\Sigma}_{yy}^{-1}\hat{\Sigma}_{zy}'$) in the simple Euclidean metric. A disadvantage of (6.22) and (6.23) is that the matrices at the left-hand side of the equations are not symmetric. Therefore, their characteristic roots and vectors are difficult to compute. A simple solution to this difficulty is to consider CC analysis as a specific form of PR analysis, hence $\hat{B}_{cc}$ is given by (see (1.16))

(6.24) $\qquad \hat{\Sigma}\hat{B}_{cc} = \hat{Q}_{cc}^{-1}\hat{B}_{cc}\hat{\Lambda}_{min,p}$

$$\text{with } \hat{Q}_{cc} = \left[\begin{array}{c|c} \hat{\Sigma}_{yy}^{-1} & 0 \\ \hline 0 & \hat{\Sigma}_{zz}^{-1} \end{array}\right].$$

Now $\hat{B}_{cc}$ can be obtained by means of the same algorithm, as the one that has been used to derive $\hat{B}_{pr}$ (see section 1), with Q replaced by $\hat{Q}_{cc}$ and with the additional assign-statement

(6.25) $\quad$ 7. $\hat{\Phi}_{min,p} := I_p - \hat{\Lambda}_{min,p}.$

The limiting distribution of $vec\hat{B}_{cc}$ can be obtained by means of the delta method. Therefore, the matrix of partial derivatives of $vecB_{cc}$ with respect to $(vec\Sigma)'$ is needed, which will be derived in the following theorem. Defining (k×1) vector $\beta_{cc,i}$ as the ith column of B_{cc} (hence $B_{cc} = [\beta_{cc,1},\ldots,\beta_{cc,p}]$), the matrix of derivatives is given by

(6.26) $\quad \nabla(B_{cc};\Sigma) = \left[\begin{array}{c} \nabla(\beta_{cc,1};\Sigma) \\ \cdot \\ \cdot \\ \cdot \\ \nabla(\beta_{cc,p};\Sigma) \end{array}\right] \qquad (kp \times k^2).$

As every vector $\beta_{cc,i}$ is in a similar way a function of Σ (they all are characteristic vectors of Σ in the metric defined by (6.3)), it is sufficient to obtain the derivative of one column of B_{cc} with respect to $(vec\Sigma)'$.

Theorem (the derivative of the CC matrix)

Let λ_{cc} and β_{cc} be a characteristic root and the corresponding unique
characteristic vector of Σ in the metric defined by Q_{cc}^{-1} (see (6.3)), that is

$$(6.27) \qquad \Sigma\beta_{cc} = Q_{cc}^{-1}\beta_{cc}\lambda_{cc}$$

with normalization

$$(6.28) \qquad \beta'_{cc,y}\Sigma_{yy}\beta_{cc,y} = 1 \quad \text{and} \quad \beta'_{cc,z}\Sigma_{zz}\beta_{cc,z} = 1.$$

Then β_{cc} may be considered as a differentiable function of Σ with vector
derivative

$$(6.29) \qquad \nabla(\beta_{cc};\Sigma) = -(\beta'_{cc} \otimes Q_{cc}^{\frac{1}{2}}(\Sigma^{*}-\lambda_{cc}I_k)^{+}Q_{cc}^{\frac{1}{2}})$$

$$+ \lambda_{cc}Q_{cc}^{\frac{1}{2}}(\Sigma^{*}-\lambda_{cc}I_k)^{+} + Q_{cc}^{\frac{1}{2}}\left[\beta'_{cc,y} \otimes \begin{bmatrix} I_m & 0 \\ \hline 0 & 0 \end{bmatrix} \vdots \beta'_{cc,z} \otimes \begin{bmatrix} 0 & 0 \\ \hline 0 & I_{k-m} \end{bmatrix}\right]$$

$$(k\times k^2).$$

The null-submatrices are defined to be of the proper dimensions, such that
they form (together with I_m and I_{k-m} respectively) square matrices of order k.
$Q_{cc}^{\frac{1}{2}}$ is a symmetric matrix of order k, such that $Q_{cc}^{\frac{1}{2}}Q_{cc}^{\frac{1}{2}} = Q_{cc}$ and
$Q_{cc}^{\frac{1}{2}}Q_{cc}^{-1}Q_{cc}^{\frac{1}{2}} = I_k$. Σ^{*} is the transformed covariance matrix defined by
$\Sigma^{*} = Q_{cc}^{\frac{1}{2}}\Sigma Q_{cc}^{\frac{1}{2}}$ and $(\Sigma^{*}-\lambda_{cc}I_k)^{+}$ denotes the Moore-Penrose inverse of $(\Sigma^{*}-\lambda_{cc}I_k)$.

Prior to the proof of this theorem, the reader should realize the similarity
between the resulting formula (6.29) and the corresponding derivative of PR
vector β_{pr} given in (1.37). The first term at the right-hand side of (6.29)
exactly equals the derivative of β_{pr} with respect to $(vec\Sigma)'$. The subscript
"cc" indicates the special form of metric matrix Q. The functional dependency
of Q_{cc} on Σ yields the second term at the right-hand side of (6.29). Because
of the many similarities between PR and CC, the following proof mainly follows

the procedure applied in section 1 in order to obtain the derivatives of β_{pr}. The only novelty is in the fact that Q is no longer assumed to be a constant, fixed matrix, which yields more complicated formulas in the derivations.

Proof

Using (6.28), expression (6.27) can be rewritten as

$$(6.30) \qquad (I_k - \tfrac{1}{2}Q_{cc}^{-1}\beta_{cc}\beta_{cc}')\Sigma\beta_{cc} = 0.$$

Taking derivatives with respect to $(vec\Sigma)'$ we obtain

$$(6.31) \qquad \nabla((I_k - \tfrac{1}{2}Q_{cc}^{-1}\beta_{cc}\beta_{cc}')\Sigma\beta_{cc};\Sigma) + \nabla((I_k - \tfrac{1}{2}Q_{cc}^{-1}\beta_{cc}\beta_{cc}')\Sigma\beta_{cc};Q_{cc}^{-1}) \; \nabla(Q_{cc}^{-1};\Sigma)$$

$$+ \nabla((I_k - \tfrac{1}{2}Q_{cc}^{-1}\beta_{cc}\beta_{cc}')\Sigma\beta_{cc};\beta_{cc}) \; \nabla(\beta_{cc};\Sigma) = 0,$$

where we should not forget that Q_{cc} as well as β_{cc} are functions of Σ. In order to obtain an explicit expression for $\nabla(\beta_{cc};\Sigma)$, it is necessary to compute the derivatives that are used as the first term, the second term and as the first factor of the last term of (6.31). Analogous to the results (1.42) and (1.44) for β_{pr} we may write

$$(6.32) \qquad \nabla((I_k - \tfrac{1}{2}Q_{cc}^{-1}\beta_{cc}\beta_{cc}')\Sigma\beta_{cc};\Sigma) = (\beta_{cc}' \otimes (I_k - \tfrac{1}{2}Q_{cc}^{-1}\beta_{cc}\beta_{cc}'))$$

$$(6.33) \qquad \nabla((I_k - \tfrac{1}{2}Q_{cc}^{-1}\beta_{cc}\beta_{cc}')\Sigma\beta_{cc};\beta_{cc}) = (\Sigma-\lambda_{cc}Q_{cc}^{-1}) - Q_{cc}^{-1}\beta_{cc}\beta_{cc}'\Sigma.$$

The partial derivatives concerning the second term of (6.31) are derived as follows. Applying the product rule for matrix differentiation (A.41), the first factor of the second term is rewritten as

$$(6.34) \qquad - \tfrac{1}{2} \frac{\partial Q_{cc}^{-1}}{\partial (vecQ_{cc}^{-1})'} \; (\beta_{cc}\beta_{cc}'\Sigma\beta_{cc} \otimes I_{k^2}).$$

Using (A.47) and replacing $\beta'_{cc}\Sigma\beta_{cc}$ by $2\lambda_{cc}$, (6.34) yields

(6.35) $\quad \nabla((I_k - \tfrac{1}{2}Q^{-1}_{cc}\beta_{cc}\beta'_{cc})\Sigma\beta_{cc};Q^{-1}_{cc}) = -((vecI_k)' \otimes I_k)(\lambda_{cc}\beta_{cc} \otimes I_{k^2})$

$$= -\lambda_{cc}((vecI_k)'(\beta_{cc} \otimes I_k) \otimes I_k)$$

$$= -\lambda_{cc}(\beta'_{cc} \otimes I_k).$$

In order to compute the second factor of the second term of (6.31), the following identity is used.

(6.36) $\quad Q^{-1}_{cc} = \begin{bmatrix} I_m & 0 \\ \hline 0 & 0 \end{bmatrix} \Sigma \begin{bmatrix} I_m & 0 \\ \hline 0 & 0 \end{bmatrix} + \begin{bmatrix} 0 & 0 \\ \hline 0 & I_{k-m} \end{bmatrix} \Sigma \begin{bmatrix} 0 & 0 \\ \hline 0 & I_{k-m} \end{bmatrix}.$

Replacing Q^{-1}_{cc} by the right-hand side of (6.36), applying vectorization property (A.11) and using matrix derivative result $\nabla(\Sigma;\Sigma) = I_{k^2}$, we obtain

(6.37) $\quad \nabla(Q^{-1}_{cc};\Sigma) = (\begin{bmatrix} I_m & 0 \\ \hline 0 & 0 \end{bmatrix} \otimes \begin{bmatrix} I_m & 0 \\ \hline 0 & 0 \end{bmatrix}) + (\begin{bmatrix} 0 & 0 \\ \hline 0 & I_{k-m} \end{bmatrix} \otimes \begin{bmatrix} 0 & 0 \\ \hline 0 & I_{k-m} \end{bmatrix}).$

Substitution of (6.32), (6.33), (6.35) and (6.37) in (6.31) yields

$(6.38) \quad (\beta'_{cc} \otimes (I_k - \frac{1}{2}Q_{cc}^{-1}\beta_{cc}\beta'_{cc}))$

$$- \lambda_{cc}(\beta'_{cc} \otimes I_k)((\begin{bmatrix} I_m & \vdots & 0 \\ \cdots & & \cdots \\ 0 & \vdots & 0 \end{bmatrix} \otimes \begin{bmatrix} I_m & \vdots & 0 \\ \cdots & & \cdots \\ 0 & \vdots & 0 \end{bmatrix}) + (\begin{bmatrix} 0 & \vdots & 0 \\ \cdots & & \cdots \\ 0 & \vdots & I_{k-m} \end{bmatrix} \otimes \begin{bmatrix} 0 & \vdots & 0 \\ \cdots & & \cdots \\ 0 & \vdots & I_{k-m} \end{bmatrix}))$$

$$+ ((\Sigma - \lambda_{cc}Q_{cc}^{-1}) - Q_{cc}^{-1}\beta_{cc}\beta'_{cc}\Sigma) \nabla(\beta_{cc};\Sigma) = 0.$$

Note the correspondence between the matrix equations (1.46) and (6.38). Except for the replacement of the subscripts "pr" by "cc", the only difference is in the fact that in the CC case we have an additional term, which is given in the second line of (6.38). From expression (1.46) the matrix $\nabla(\beta_{pr};\Sigma)$ of partial derivatives in the PR case has been derived. Application of corresponding techniques with respect to formula (6.38) we obtain

$$(6.39) \quad \nabla(\beta_{cc};\Sigma) = - (\beta'_{cc} \otimes Q_{cc}^{\frac{1}{2}}(\Sigma^* - \lambda_{cc}I_k)^+ Q_{cc}^{\frac{1}{2}})$$

$$+ \lambda_{cc}Q_{cc}^{\frac{1}{2}}(\Sigma^* - \lambda_{cc}I_k)^+ Q_{cc}^{\frac{1}{2}}(\beta'_{cc} \otimes I_k) \times$$

$$((\begin{bmatrix} I_m & \vdots & 0 \\ \cdots & & \cdots \\ 0 & \vdots & 0 \end{bmatrix} \otimes \begin{bmatrix} I_m & \vdots & 0 \\ \cdots & & \cdots \\ 0 & \vdots & 0 \end{bmatrix}) + (\begin{bmatrix} 0 & \vdots & 0 \\ \cdots & & \cdots \\ 0 & \vdots & I_{k-m} \end{bmatrix} \otimes \begin{bmatrix} 0 & \vdots & 0 \\ \cdots & & \cdots \\ 0 & \vdots & I_{k-m} \end{bmatrix})).$$

Using some elementary Kronecker product properties this term can be rewritten, such that we obtain the result as proposed in (6.29). ◇

When $\hat{\Sigma}$ is defined to be an asymptotically normally distributed estimator of covariance matrix Σ, then the limiting distribution of $\hat{\beta}_{cc}$ (the sample analogue of β_{cc}) is given by

$$(6.40) \quad \sqrt{N}(\hat{\beta}_{cc} - \beta_{cc}) \xrightarrow{D} N(0, [\nabla(\beta_{cc};\Sigma)]V[\nabla(\beta_{cc};\Sigma)]')$$

with V the asymptotic covariance matrix of $\sqrt{N}(\text{vec}\hat{\Sigma} - \text{vec}\Sigma)$ and $\nabla(\beta_{cc};\Sigma)$ the matrix of partial derivatives, that has been obtained in the preceding theorem. So far, most attention has been given to the estimation of the elements of B_{cc}, resulting in a limiting distribution, denoted in (6.40). However, the method of moments approach does not only yield asymptotically normally distributed estimators of the columns of B_{cc} but also consistent estimators of the canonical correlations. The sample canonical correlations are given by the p diagonal elements of $\hat{\phi}_{max,p}$ (see (6.22) and (6.23)). Let the canonical correlations be denoted by $\phi_1, \phi_2, \ldots, \phi_p$ with

(6.41) $\phi_1 \geq \phi_2 \geq \ldots \geq \phi_p$ and $p \leq \min(m,(k-m))$.

It is known from (6.16) that ϕ_i ($1 \leq i \leq p$) denotes the covariance between the ith pair of canonical variables. It is important to know, which canonical correlations are very small ("almost zero"), as this can lead to substantial reduction in dimensionality. For instance when the last $\min(m,(k-m))-p$ canonical correlations are (almost) zero and $\phi_p > 0$, then the relationship between Y and Z can by summarized by means of the first p canonical variables. One of the most important procedures in canonical correlation analysis is to test the hypothesis that the smallest canonical correlation coefficients are zero. For this purpose, the asymptotic distributions of the sample canonical correlations $\hat{\phi}_i$ are needed. Asymptotic joint and marginal distributions of $\hat{\phi}_i$ (i=1,...,p) have been considered by Hsu (1941) and Lawley (1959), based on multivariate normal distributed sample observations. More recently Glynn and Muirhead (1978) derived asymptotic expansions for these limiting distributions. Muirhead and Waternaux (1980) considered asymptotic theory and tests for canonical correlation analysis in case of elliptical distributed sample observations with finite fourth-order moments. In this study we shall return to the asymptotic distribution of the canonical correlations (and in general to characteristic roots of random positive definite, symmetric matrices) in chapter V. In the existing statistical literature, less attention has been given to the limiting distribution of the columns of $\hat{B}_{cc}$, that yield the canonical variables. Recently Tayler (1981) applied general results of asymptotic inference for characteristic vectors to canonical correlation analysis, where the sample is drawn for an elliptical distribution.

As usual, this section ends by specifying the limiting distribution given in

(6.40), for a so-called ideal sample consisting of N i.i.d. observations. The ideal sample situation has already been studied in the preceding sections in the context of other multivariate statistics. For clearness' sake, the form of the ideal sample V (obtained by means of the central limit theorem) is here repeated

(6.42) $\quad V = E(vec(XX'-\Sigma))(vec(XX'-\Sigma))'.$

Theorem ($\hat{\beta}_{cc}$ for random i.i.d. sample observations)
Let $\{X_n: n=1,\ldots,N\}$ be a sample of N i.i.d. random observations on X and let $\hat{\beta}_{cc}$ be the MME of β_{cc} defined in (6.27), with normalization as denoted in (6.28). Then

(6.43) $\quad \sqrt{N}(\hat{\beta}_{cc} - \beta_{cc}) \overset{D}{\to}$

$$N(0, \ Q_{cc}^{\frac{1}{2}}(\Sigma^* - \lambda_{cc} I_k)^+ Q_{cc}^{\frac{1}{2}}$$

$$\times E(XX' + \lambda_{cc} \begin{bmatrix} YY' & 0 \\ \hline 0 & ZZ' \end{bmatrix}) \beta_{cc} \beta_{cc}' (XX' + \lambda_{cc} \begin{bmatrix} YY' & 0 \\ \hline 0 & \dot{Z}Z' \end{bmatrix})$$

$$\times Q_{cc}^{\frac{1}{2}}(\Sigma^* - \lambda_{cc} I_k)^+ Q_{cc}^{\frac{1}{2}}).$$

Comparison of (6.43) with (1.51) yields the conclusion, that the limiting distribution of $\hat{\beta}_{cc}$ resembles the limiting distribution of $\hat{\beta}_{pr}$ to a large extent. The difference is in the fact, that in the asymptotic covariance matrix the factor XX' is replaced (twice) by

$(XX' + \lambda_{cc} \begin{bmatrix} YY' & 0 \\ \hline 0 & ZZ' \end{bmatrix})$, which is caused by the randomness of the metric matrix $\hat{Q}_{cc}.$

Proof
The asymptotic normality of $\hat{\beta}_{cc}$ has already been denoted in (6.40). The covariance matrix of the limiting distribution can be computed by substituting

(6.29) and (6.42) in the matrix product at the right-hand side of (6.40). Let us start by considering the pre-multiplication of $\text{vec}(XX'-\Sigma)$, which is the first factor of the middle matrix V, by the first term of the gradient matrix. Application of (A.11) yields

$$(6.44) \quad -(\beta'_{cc} \otimes Q^{\frac{1}{2}}_{cc}(\Sigma^* - \lambda_{cc}I_k)^+ Q^{\frac{1}{2}}_{cc})\text{vec}(XX'-\Sigma) = -Q^{\frac{1}{2}}_{cc}(\Sigma^* - \lambda_{cc}I_k)^+ Q^{\frac{1}{2}}_{cc}XX'\beta_{cc}.$$

In a similar way we obtain by pre-multiplication of $\text{vec}(XX'-\Sigma)$ by the second term of the gradient matrix

$$(6.45) \quad \lambda_{cc}Q^{\frac{1}{2}}_{cc}(\Sigma^* - \lambda_{cc}I_k)^+ Q^{\frac{1}{2}}_{cc}(Q^{-1}_{cc} - \begin{bmatrix} YY' & \vdots & 0 \\ --- & \vdots & --- \\ 0 & \vdots & ZZ' \end{bmatrix}) \, \beta_{cc}$$

$$= -\lambda_{cc}Q^{\frac{1}{2}}_{cc}(\Sigma^* - \lambda_{cc}I_k)^+ Q^{\frac{1}{2}}_{cc} \begin{bmatrix} YY' & \vdots & 0 \\ --- & \vdots & --- \\ 0 & \vdots & ZZ' \end{bmatrix} \, \beta_{cc}.$$

Summation of (6.44) and (6.45) yields an expression for $[\nabla(\beta_{cc};\Sigma)]\text{vec}(XX'-\Sigma)$. From the definition of V in (6.42) it follows, that an expression for the asymptotic covariance function of $\sqrt{N}(\hat{\beta}_{cc} - \beta_{cc})$ is attained (as proposed in (6.43)) when $[\nabla(\beta_{cc};\Sigma)]\text{vec}(XX'-\Sigma)$ is post-multiplied by its own transpose. ◇

With this section we end the chapter on principal relations. In the sections 3 to 6 some special types of PR have been analysed, that were defined by selecting a specific matrix Q and placing a priori restrictions on parameter matrix B. Q defines a metric in which the distance minimization takes place (the projection direction) and B defines the subspace on which the vector X is projected. The connections between PR analysis and the linear models like simultaneous equations systems and regression equations is straightforward, as they all describe linear relations for the observation vector X_n. However, the connection between PR analysis and CC analysis is less transparent, as CC analysis is usually looked upon as a generalization of multiple correlation analysis. In this section it was shown that the intuitively clear replacement

of "linear compounds B'_yY and B'_zZ of Y and Z with maximum correlation" by
"linear compounds of $\tilde{Y}$ and $\tilde{Z}$, such that $B'_y\tilde{Y} = B'_z\tilde{Z}$ with $\tilde{Y}$ and $\tilde{Z}$ as close as
possible to Y and Z", yields a minimum distance approach of CC analysis.
Comparing this approach with regression analysis, CC analysis may be regarded
as a technique that "explains" a linear combination of Y variables by a linear
combination of Z variables. This interpretation of CC analysis reminds us to
simultaneous equations systems. From the minimum distance specification it
follows indeed that SE and CC analysis are closely related. The difference is
in the fact that diagonal submatrices of Q, referring to the Y and Z
dimensions of X, are differently specified.

Although the end of the chapter on PR analysis has been reached, this does not
mean that we could not think of any other assumptions on Q and B, that would
yield attractive and readily interpretive parameters. For instance one may
think of setting the metric matrix equal to

$$(6.46) \qquad Q = \left[\begin{array}{c|c} \Sigma_{yy}^{-1} & 0 \\ \hline 0 & I_{k-m} \end{array} \right],$$

yielding a statistical technique that can be seen as an intermediate between
SUR and CC analysis, which is in the statistical literature known under the
name "redundancy analysis". This technique has been introduced by Van den
Wollenberg (1977). Extensions of this theory have been given by Johansson
(1981) and Tyler (1982) and its connections with CC and SUR have been studied
by Muller (1981).

Another point of interest in the statistical literature is to describe the
random vector X by a random vector of smaller dimension, consisting of latent
variables, that has inherent statistical properties. In the next chapter this
problem will be considered in the minimum distance context.

IV. Principal Factors

IV.1 Basic formulation of principal factors

Similar to the preceding chapter, the random vector of interest will be
denoted by a (k×1) vector X and it is assumed that X has zero mean. At the
cost of being redundant, some of the distributional assumptions and notational
conventions will be repeated. It is assumed that we do not have any knowledge
about the distribution of X, except for the existence of second and fourth-
order moments. This means that the covariance matrix

$$(1.1) \qquad \Sigma = E(XX') \qquad\qquad\qquad (k×k)$$

of X and the asymptotic covariance matrix

$$(1.2) \qquad V = E(vec(XX'))(vec(XX'))' - (vec\Sigma)(vec\Sigma)' \qquad (k^2×k^2)$$

of vec(XX') are bounded. Boundedness of Σ is required as the purpose of this
study is to derive the model parameters of interest as (differentiable)
functions of Σ. Because of the assumption of existing fourth-order moments it
is possible to make statistical large sample inferences about the
estimator $\hat{\Sigma}$. By means of the delta method also the limiting distributions of
differentiable functions of $\hat{\Sigma}$ can be obtained.

In chapter III we considered how to estimate a (k×p) parameter matrix B, that
defines the p so-called principal relations which are approximately fulfilled
by X. In fact the relations are exactly fulfilled by $\tilde{X}$, which is the Q-
orthogonal projection of X onto $\{\xi: B'\xi=0\}$. The optimal matrix B_{pr} appeared to
be a function of Σ and Q and defines a linear (k-p)-dimensional subspace of
$\mathbb{R}^k$, that fits the distribution of X "as accurately as possible". This means
that the random vector $\tilde{X}$ (and therefore approximately the vector X) can be
interpreted as a linear transformation of a random non-observable (k-p)-
dimensional vector. This property will be the starting point of the study of
the linear structural models in this chapter. Again the purpose is to describe
a linear subspace that fits the distribution of X as accurately as possible.
However, this will not be done (like in PR analysis) by defining the p
vectors, denoted by the columns of B_{pr}, orthogonal to the optimal (k-p)-
dimensional linear subspace. Now we shall look for a so-called parametric
equation of the subspace. What is meant by the parametric equation of a

subspace? A q-dimensional, say, linear subspace of R^k through the origin can be described as the linear combination of a set of q independent (k×1) vectors α_i (i=1,...,q). That is, the subspace can be denoted by

(1.3) $\xi = \upsilon_1\alpha_1 + \upsilon_2\alpha_2 + \ldots + \upsilon_q\alpha_q$

with $\upsilon_1,\ldots,\upsilon_q$ the (scalar) coefficients of the linear combination, representing the coordinates of ξ with respect to the basis $\{\alpha_i : i=1,\ldots,q\}$ for the q dimensional linear subspace. In matrix notation (1.3) is given by

(1.4) $\xi = A\upsilon$

with $A = [\alpha_1,\ldots,\alpha_q]$ a (k×q) matrix and $\upsilon = (\upsilon_1,\ldots,\upsilon_q)'$ a (q×1) vector. As a result, all vectors defined by (1.4) together form the q-dimensional linear subspace spanned by the columns of A, which is denoted by span(A). Defining corresponding relations for random vectors, principal factor (PF) analysis can be defined as the compound of two minimization problems. Firstly, a vector $\hat{X}$ must be derived, such that $\hat{X}$ is at a minimum Q-distance of X with $\hat{X} \in$ span(A). Hence, if ξ is an auxiliary vector in span(A), then $\hat{X}$ is defined to be the value of ξ that minimizes the (squared) Q-distance between X and ξ. As $\xi \in$ span(A), there exists a (q×1) vector υ such that (1.4) holds and in mathematical terminology the minimum distance problem yields the following mathematical programming problem

(1.5) $\min_{\substack{\xi \\ \xi=A\upsilon}} d_Q^2(X,\xi) = \min_{\substack{\xi \\ \xi=A\upsilon}} (X-\xi)'Q(X-\xi),$

or equivalently

(1.6) $\min_{\upsilon} (X-A\upsilon)'Q(X-A\upsilon).$

The optimal value for υ, say U, is obtained as a function of X and A. The optimal value $\hat{X}$ of ξ is in a natural way obtained by

(1.7) $\hat{X} = AU.$

The second objective is to find the linear subspace span(A), such that the expected squared distance between X and $\hat{X}$ (given by the solution of (1.5)) is minimal. This subspace is spanned by the columns of matrix A_{pf}, say, that is obtained as the optimal solution of the mathematical programming problem given by

(1.8) $\min_{A} E(X-\hat{X})'Q(X-\hat{X}).$

Expressions (1.6) and (1.8) together form the basic formulation of the PF problem, given by

(1.9) $\min_{A} E[\min_{\upsilon} (X-A\upsilon)'Q(X-A\upsilon)].$

The solution of (1.9) is given in the following theorem.

Theorem (the PF matrix)
Let the mathematical programming problem be defined by (1.9) with

X a (k×1) random vector
υ a (q×1) auxiliary vector
A a (k×q) parameter matrix of rank q
Q a (k×k) positive definite symmetric matrix.

The optimal value U of υ is then given by

(1.10) $U = (A'QA)^{-1}A'QX$

and the optimal value A_{pf} of A is defined by

(1.11) $\Sigma Q A_{pf} = A_{pf}\Lambda_{max,q}$

with Σ the covariance matrix of X and $\Lambda_{max,q}$ the (q×q) diagonal matrix with on its diagonal the q largest characteristic roots of Σ in the metric defined by Q^{-1}.

Combination of (1.10) and (1.7) yields

(1.12) $\hat{X} = A(A'QA)^{-1}A'QX.$

From (1.11) it follows that $\hat{X}$ is the Q-orthogonal projection of X on span(A). The best fitting linear subspace is spanned by the columns of A_{pf} defined in (1.11). Rewriting (1.11) as

(1.13) $\Sigma(QA_{pf}) = Q^{-1}(QA_{pf})\Lambda_{max,q}$

it can be seen that the columns of QA_{pf} are the q characteristic vectors of Σ corresponding with largest characteristic roots in the Q^{-1}-metric.

Proof

Setting the derivative of the objective function of the inner minimization problem in (1.9) with respect to υ equal to zero, we obtain

(1.14) $-2A'QX + 2 A'QAU = 0.$

As Q is positive definite and A is of rank q, with $q \le k$, the matrix A'QA is positive definite as well, and therefore non-singular. Pre-multiplication of (1.14) by $(A'QA)^{-1}$ yields (1.10). Substitution of the resulting optimal value $\hat{X}$ = AU (written out in (1.12)) in the outer minimization problem (see (1.8)), results in

(1.15) $\min_{A} E(X'(Q - QA(A'QA)^{-1}A'Q)X).$

Using the trace operator and taking expectations, (1.15) can be rewritten as

(1.16) $\min_{A} tr((Q - QA(A'QA)^{-1}Q)A'Q)\Sigma),$

the solution of which equals the optimum value of

(1.17) $\max_{A} tr(QA(A'QA)^{-1}A'Q\Sigma).$

Notice the similarities between the minimization problem in PR analysis (see (III.1.23)) and the maximization problem defined in (1.17). The derivative of the objective function in (1.17) with respect to A is given in Balestra (1976, p. 74). Setting the derivative equal to zero, the optimal value A_{pf} of A is obtained as a solution of the following matrix equation

$$(1.18) \qquad (I_k - QA_{pf}(A'_{pf}QA_{pf})^{-1}A'_{pf})Q\Sigma QA_{pf}(A'_{pf}QA_{pf})^{-1} = 0.$$

Just like in the PR case, the resulting system of equations seems a little obscure, as this matrix equation is a rather complex function of the parameter matrix A_{pf}. In order to get an idea of a possible solution, we consider formula (1.18) for a certain class of possible A_{pf} matrices. Let us for instance consider the matrix product QA_{pf} and restrict its possible values to the class of orthonormal matrices in the Q^{-1}-metric. That is, we assume $(QA_{pf})'Q^{-1}(QA_{pf}) = A'_{pf}QA_{pf} = I_q$. Then (1.18) reduces to

$$(1.19) \qquad A'_{pf}QA_{pf} = I_q \;\Rightarrow\; (Q - (QA_{pf})(QA_{pf})')\Sigma(QA_{pf}) = 0.$$

The right-hand side of (1.19) is known to be solved by the characteristic vectors of Σ in the Q^{-1}-metric, or, more formally, the stationary value of A_{pf} is defined by the system of equations

$$(1.20) \qquad \Sigma(QA_{pf}) = Q^{-1}(QA_{pf})\Lambda_Q = A_{pf}\Lambda_Q$$

with Λ_q a diagonal matrix with q of the k characteristic roots of Σ in the Q^{-1}-metric on its diagonal.

Now we return to our original equation (1.18). Note that also without the orthonormality assumption, (1.20) defines a solution of (1.18), which may be checked by substitution of $A_{pf}\Lambda_q$ for ΣQA_{pf} in (1.18) and verification of the equality sign. From substitution of (1.20) in (1.17) it follows that the stationary values of (1.17) are given by $tr(\Lambda_q)$ and as a consequence a maximum is reached when the diagonal elements of Λ_q equal the q maximal characteristic roots of Σ in the Q^{-1}-metric. As a result we may conclude that QA_{pf} represents the characteristic vectors of Σ, corresponding with the q largest characteristic roots in the Q^{-1}-metric. ◊

From (1.11) the derivation of the columns of A_{pf} follows in a natural way. They are obtained by pre-multiplication of the characteristic vectors of Σ in the Q^{-1}-metric, corresponding with the q largest characteristic roots, by Q^{-1}. Comparing A_{pf} with B_{pr} of section III.1., many similarities are to be noticed between the two minimum distance problems and their solutions. For instance in both techniques, an important role is played by the characteristic roots and vectors of Σ. More about these differences and similarities will be discussed in section IV.2.

As mentioned before, the characteristic vectors do not yield a unique solution of the minimization problem. A natural way of defining a single solution is to normalize the columns of QA_{pf}, so that their lenghts equal a fixed, predetermined value. Two important normalizations are (together with the orthogonality property of the columns of QA_{pf}) given by

$$(1.21) \qquad \|QA_{pf}\|^2_{Q^{-1}} = I_q \quad \Rightarrow \quad \|A_{pf}\|^2_Q = A'_{pf}QA_{pf} = I_q$$

or

$$(1.22) \qquad \|QA_{pf}\|^2_{Q^{-1}} = \Lambda_{max,q} \quad \Rightarrow \quad \|A_{pf}\|^2_Q = A'_{pf}QA_{pf} = \Lambda_{max,q}.$$

Indirectly, the normalization of QA_{pf} defines the covariance matrix of the random vector U. Using (1.10) it follows that the expectation of U equals zero, and its covariance matrix is given by

$$(1.23) \qquad E(UU') = (A'_{pf}QA_{pf})^{-1}A'_{pf}Q\Sigma QA_{pf}(A'_{pf}QA_{pf})^{-1}.$$

Replacing ΣQA_{pf} by $A_{pf}\Lambda_{max,q}$, (1.23) can be rewritten as

$$(1.24) \qquad E(UU') = \Lambda_{max,q}(A'_{pf}QA_{pf})^{-1}.$$

Hence the unit normalization and the characteristic root normalization, defined in (1.21) and (1.22) respectively, yield the following implications

$$(1.25) \qquad \|A_{pf}\|^2_Q = I_q \quad \Rightarrow \quad E(UU') = \Lambda_{max,q}$$

and

$$(1.26) \qquad \|A_{pf}\|^2_Q = \Lambda_{max,q} \quad \Rightarrow \quad E(UU') = I_q.$$

Naturally the normalization of A_{pf} also partly determines the covariance between the vector X and the vector U. As (according to (1.10))

(1.27) $E(XU') = \Sigma Q A_{pf}(A'_{pf} Q A_{pf})^{-1}$,

we may write, after substitution of $\Sigma Q A_{pf}$ by $A_{pf}\Lambda_{max,q}$, analogous to (1.25) and (1.26)

(1.28) $\|A_{pf}\|_Q^2 = I_q$ $\Rightarrow$ $E(XU') = A_{pf}\Lambda_{max,q}$

and

(1.29) $\|A_{pf}\|_Q^2 = \Lambda_{max,q}$ $\Rightarrow$ $E(XU') = A_{pf}$.

Why the name "principal factors" for this technique? In classical statistical literature and especially psychometric literature, a somewhat similar approach is chosen in factor analysis. In contrast to the PF analysis, where no linear model for X is assumed, the X variables are in factor analysis specified ab initio to be linear combinations of a smaller (than k) number of independent unobservables plus an independent error term. Using the same notation as practiced in the preceding theorem on the PF matrix, the factor analysis model can be described by

(1.30) $X = AU + \varepsilon$

with ε a (k×1) vector of latent random disturbance terms which are assumed to be mutually independent. This linear model represents an explanation of the covariance structure of X. The components of U (usually called the "common factors") generate the covariances among the components of X, while the components of ε (called the "specific factors") contribute only to the variances of their particular parts of X, as they are assumed to be mutually uncorrelated. The matrix A consists of parameters α_{ij} (i=1,...,k and j=1,...,q) that reflect the importance of the jth common factor in the composition of the ith observable variable. Hence comparing the linear relation $X = A_{pf}U$ of PF analysis with relation (1.30), it is obvious that in classical factor analysis one is interested in a full "explanation" of X in terms of the unobservable factors U, while PF analysis considers the linear combination $A_{pf}U$ that defines the q-dimensional (with q a priori chosen) linear subspace with minimum distance to X. Hence $A_{pf}U$ does not necessarily need to specify the covariance structure of X completely. As in PF analysis the number of common factors is restricted to q, the factor model need not to be complete, as k-q dimensions are "neglected". It is for this reason that in this study the name "factor analysis" is extended to "principal factor analysis". The q components of U will be called the principal factors of X. Similar to factor analysis the elements of matrix A_{pf} are referred to as factor loadings. Because of the unobservability of U, the usefulness of factor analysis in empirical investigations largely depends on the ability of identifying these factors with intuitively interpretable entities. The theory of factor analysis has a long historical background. The mathematical model for factor structures was introduced by Spearman (1904). He hypothesized that the correlations among the set of intelligence-test scores could be generated by a single common factor that stands for the general intelligence of the respondent and a set of specific factors reflecting the qualities per individual test. Note that this model deals with only one common factor. Late this model was extended to include more than one common factor (see for instance Thurstone (1945)). Many different techniques are proposed for extracting the factor coefficients from the covariance matrix of X. Various approaches have for instance been discussed by Harman (1967), who also gave a bibliography of more than 500 references. Lawley (1940, 1942, 1943) described the extraction of factor parameters as a maximum likelihood technique, which is the most widely known estimation method for the factor loadings, described in many introductory statistical textbooks. (See for instance Van de Geer (1971, chapter 13) and Morrison (1978, chapter 9)). In order to apply the maximum likelihood technique it is common practice to assume that the common factor variates in U and the specific factors in ε are independently and normally distributed with zero means. Furthermore the common factors are assumed to have unit variances and the variances of the specific factors all equal σ_ε^2. Then from (1.30) it follows that

$$(1.31) \qquad \Sigma = AA' + \sigma_\varepsilon^2 I_k,$$

which clearly represents the decomposition of Σ into a part generated by the common factors and a part of the variances contributed by the disturbance te ε. The logarithm of the likelihood can be obtained in terms of the Wishart density for the sample analogue $\hat{\Sigma}$ of Σ, if X is normally distributed. It can be shown that the maximum likelihood estimator of the columns of A are given by the characteristic vectors of $\hat{\Sigma}$, corresponding with the q maximum characteristic roots $\hat{\lambda}_i$ ($i=1,\ldots,q$), in the I_k-metric. The maximum likelihood estimator of σ_ε^2 is obtained by means of the q maximum characteristic roots and equals

$$(1.32) \qquad \hat{\sigma}_\varepsilon^2 = \frac{\mathrm{tr}\hat{\Sigma} - \sum\limits_{i=1}^{q} \hat{\lambda}_i}{k-q}$$

(for details of the derivations the reader is referred to Lawley and Maxwell (1960)).

Let us return to the matrix A_{pf} of PF analysis, which is obtained as a solution of a minimum distance problem. Recalling (1.26), it is clear that $\|A_{pf}\|_Q^2 = \Lambda_{max,q}$ is the most natural normalization choice for the columns of A_{pf}, as the covariance matrix of U then equals I_k, which corresponds with the covariance structure assumed in the classical form of factor analysis. Another attractive property of the characteristic root normalization is that the $(i,j)^{th}$ entry of A_{pf} denotes the covariance between X_i (the ith component of X) and U_j (the jth component of U), as shown in (1.29). So far, the PF technique has been analysed in the metric defined by Q, yielding characteristic vectors of Σ in the Q^{-1}-metric as optimal solutions. However, in practice this solution is hard to work with, as the fact that we are not working in the well-known simple Euclidean I_k-metric is a computational disadvantage. Therefore the transformation of vector X is proposed, that has already been applied in section III.1. to evade corresponding difficulties. In section III.1. it has been shown that, if Q is a symmetric and positive definite matrix of order k, then there exists a symmetric (k×k) matrix $Q^{\frac{1}{2}}$ such that

$$(1.33) \qquad Q^{\frac{1}{2}}Q^{\frac{1}{2}} = Q \quad \text{and} \quad Q^{\frac{1}{2}}Q^{-1}Q^{\frac{1}{2}} = I_k$$

(see also (III.1.11). If X^* is the transformation of X defined by

$$(1.34) \qquad X^* = Q^{\frac{1}{2}}X$$

with $Q^{\frac{1}{2}}$ in conformity with (1.33), then the covariance matrix of X^* (denoted by Σ^*) is given by

(1.35) $\Sigma^* = Q^{\frac{1}{2}} \Sigma Q^{\frac{1}{2}}$

and the Q-distance between X and ξ equals the I_k-distance between X^* and $\xi^* = Q^{\frac{1}{2}}\xi$. Hence the PF analysis for X^* is defined by the minimization of $(X^* - \xi^*)'(X^* - \xi^*)$, yielding solutions

(1.36) $\hat{X}^* = A^* U^* = A^*(A^{*\prime}A^*)^{-1}A^{*\prime}X^*$

and

(1.37) $\Sigma^* A_{pf}^* = A_{pf}^* \Lambda_{max,q}^*.$

Combining (1.37) and its untransformed counterpart $\Sigma A_{pf} = Q^{-1}A_{pf}\Lambda_{max,q}$, it follows that

(1.38) $A^* = Q^{\frac{1}{2}}A$ (or $A = Q^{-\frac{1}{2}}A^*$) and $\Lambda_{max,q} = \Lambda_{max,q}^*.$

Substitution of (1.38) in $U^* = (A^{*\prime}A^*)^{-1}A^{*\prime}X^*$ yields

(1.39) $U = U^*$ and $\hat{X} = Q^{-\frac{1}{2}}\hat{X}^*.$

Consequently, the optimal values A_{pf} and $\Lambda_{max,q}$ are easily obtained from the characteristic roots $\Lambda_{max,q}^*$ and the characteristic vectors in A_{pf}^* of covariance matrix Σ^*. Normalization of the columns of A_{pf}^* in the I_k-metric corresponds with normalization of A_{pf} in the Q-metric, as $A_{pf}^{*\prime}A_{pf}^* = A_{pf}'QA_{pf}$. Hence working with the transformed vector X^*, it seems a natural choice to define a unique solution of A_{pf}^* with normalization restriction

(1.40) $A_{pf}^{*\prime}A_{pf}^* = \Lambda_{max,q}.$

In order to obtain a consistent estimator of A_{pf} it is assumed that we have knowledge of a consistent estimator $\hat{\Sigma}$ of Σ. As the parameter matrix A_{pf} is obtained as a continuous (implicit) function of Σ, a consistent estimator $\hat{A}_{pf}$ of A_{pf} is obtained by replacing in (1.11) the covariance matrix Σ by its consistent estimator $\hat{\Sigma}$, yielding

$$(1.41) \qquad \hat{\Sigma}Q\hat{A}_{pf} = \hat{A}_{pf}\hat{\Lambda}_{max,q}.$$

Application of the factorization of Q, which has been dealt with in section III.1, yields the following algorithm to compute $\hat{A}_{pf}$

1. compute $\hat{\Sigma}$ (consistent estimator of Σ);
2. deduce from Q its characteristic roots (in diagonal matrix Δ) and its characteristic vectors (K);
3. $Q^{\frac{1}{2}} := K\Delta^{\frac{1}{2}}K'$;
4. $\hat{\Sigma}^* := Q^{\frac{1}{2}}\hat{\Sigma}Q^{\frac{1}{2}}$;
5. deduce from $\hat{\Sigma}^*$ its q maximal characteristic roots and the corresponding characteristic vectors, yielding $\hat{\Lambda}_{max,q}$ and $\hat{A}^*_{pf}$;
6. $\hat{A}_{pf} := Q^{-\frac{1}{2}}\hat{A}^*_{pf}$.

If $\hat{\Sigma}$ is not only a consistent estimtor of Σ, but also asymptotically normally distributed, that is

$$(1.42) \qquad \sqrt{N}(\text{vec}\hat{\Sigma} - \text{vec}\Sigma) \overset{D}{\to} N(0, V),$$

then the limiting distribution of $\text{vec}\hat{A}_{pf}$ can be obtained by means of the delta method. Therefore the matrix of partial derivatives of $\text{vec}A_{pf}$, with respect to $\text{vec}\Sigma$ is needed. Like in the preceding sections, the results will be given for one column α_{pf} of A_{pf} only, as all columns are in a similar way a function of Σ (viz. characteristic vectors in the Q^{-1}-metric pre-multiplied by Q^{-1}).

Theorem (the derivatives of the PF matrix)
Let λ and $Q\alpha_{pf}$ be a characteristic root and the corresponding characteristic vector of Σ in the Q^{-1}-metric with characteristic root normalization:

$$(1.43) \qquad \Sigma Q\alpha_{pf} = \alpha_{pf}\lambda , \quad \alpha'_{pf}Q\alpha_{pf} = \lambda.$$

Then α_{pf} may be considered as a differentiable function of Σ with

$$(1.44) \qquad \nabla(\alpha_{pf};\Sigma) = -Q^{-1}(\alpha'_{pf}Q \otimes Q^{\frac{1}{2}}(\Sigma^{*}-\lambda I_{k})^{+}Q^{\frac{1}{2}}) + \tfrac{1}{2}\tfrac{1}{\lambda}\alpha_{pf}(\alpha'_{pf}Q \otimes \alpha'_{pf}Q)$$

with $Q^{\frac{1}{2}}$ the transformation matrix defined in (1.33), $\Sigma^{*} = Q^{\frac{1}{2}}\Sigma Q^{\frac{1}{2}}$ and $(\Sigma^{*}-\lambda I_{k})^{+}$ the Moore-Penrose inverse of $(\Sigma^{*}-\lambda I_{k})$.

Naturally the resulting derivative given in (1.44) corresponds for a great part with the derivative of the principal relation vector β_{pr}, which was in section III.1 shown to be $-(\beta'_{pr} \otimes Q^{\frac{1}{2}}(\Sigma^{*}-\lambda I_{k})^{+}Q^{\frac{1}{2}})$ (see (III.1.37)), as we are in both cases dealing with characteristic vectors. In the PR case the characteristic vector was denoted by β_{pr} and here the characteristic vector equals $Q\alpha_{pf}$. The difference between $\nabla(\beta_{pr};\Sigma)$ and the first term at the right-hand side of $\nabla(\alpha_{pf};\Sigma)$ is in the first factor of the Kronecker product, where β_{pr} is replaced by $Q\alpha_{pf}$. The pre-multiplication of the Kronecker product by Q^{-1} is caused by the fact that the characteristic vector of Σ is pre-multiplied by Q^{-1} in order to get α_{pf}. So far, everything is clear and (1.44) seems to follow directly from the theorem on the derivative of β_{pr}, proven in section III.1. However, this time the length of the characteristic vector does not equal a constant. In the PR analysis we have $\beta'_{pr}Q^{-1}\beta_{pr} = 1$, while in the PF analysis the length of $Q\alpha_{pf}$ is hypothesized to equal the corresponding characteristic value. This difference in normalization causes an additional term for the gradient matrix of α_{pf}, which is given by the second term at the right-hand side of (1.44). A formal derivation is given in the following proof, that mainly follows the line of the proof for the derivatives of β_{pr}.

Proof

Replacing λ by $\alpha'_{pf}Q\alpha_{pf}$, the first equation of (1.43) can be rewritten as

$$(1.45) \qquad (\Sigma - \alpha_{pf}\alpha'_{pf})Q\alpha_{pf} = 0.$$

Taking the derivative of (1.45) with respect to $(vec\Sigma)'$ and applying the chain-rule we obtain

$$(1.46) \qquad \nabla((\Sigma-\alpha_{pf}\alpha'_{pf})Q\alpha_{pf};\Sigma) + \nabla((\Sigma-\alpha_{pf}\alpha'_{pf})Q\alpha_{pf};\alpha_{pf}) \nabla(\alpha_{pf};\Sigma) = 0$$

with the second derivative of the second term denoting the matrix of interest. Using similar techniques as applied in the aforementioned proof for β_{pr}, the remaining two derivatives can be worked out, yielding

(1.47) $\quad (\alpha'_{pf}Q \otimes I_k) + ((\Sigma Q-\lambda I_k) - 2\alpha_{pf}\alpha'_{pf}Q)\, \nabla(\alpha_{pf};\Sigma) = 0.$

Rewriting (1.47) we obtain

(1.48) $\quad (\Sigma-\lambda Q^{-1})Q\, \nabla(\alpha_{pf};\Sigma) = -(\alpha'_{pf}Q \otimes I_k) + 2\alpha_{pf}\alpha'_{pf}Q\, \nabla(\alpha_{pf};\Sigma).$

Using the transformation matrix $Q^{\frac{1}{2}}$, that has been defined in (1.33), Σ and Q^{-1} can be rewritten as $Q^{-\frac{1}{2}}\Sigma^*Q^{-\frac{1}{2}}$ and $Q^{-\frac{1}{2}}Q^{-\frac{1}{2}}$ respectively, yielding

(1.49) $\quad \nabla(\alpha_{pf};\Sigma) = -Q^{-1}Q^{\frac{1}{2}}(\Sigma^*-\lambda I_k)^+Q(\alpha'_{pf}Q \otimes I_k)$

$\quad\quad\quad + 2Q^{-1}Q^{\frac{1}{2}}(\Sigma^*-\lambda I_k)^+Q^{\frac{1}{2}}\alpha_{pf}\alpha'_{pf}Q\, \nabla(\alpha_{pf};\Sigma)$

$\quad\quad\quad + Q^{-1}Q^{\frac{1}{2}}(I_k- (\Sigma^*-\lambda I_k)^+(\Sigma^*-\lambda I_k))U$

with U an arbitrary $(k \times k^2)$ matrix. The second term at the right-hand side of (1.49) contains a factor $(\Sigma^*-\lambda I_k)^+Q^{\frac{1}{2}}\alpha_{pf} = (\Sigma^*-\lambda I_k)^+\alpha^*_{pf}$. As according to (1.37) $(\Sigma^*-\lambda I_k)\alpha^*_{pf} = 0$, it follows from (A.25) that also $(\Sigma^*-\lambda I_k)^+\alpha^*_{pf} = 0$, and consequently the complete second term at the right-hand side of (1.49) equals zero. Because of the characteristic root normalization (1.43) we have

(1.50) $\quad 2\alpha'_{pf}Q\, \nabla(\alpha_{pf};\Sigma) = \nabla(\lambda;\Sigma),$

and combination of (1.49) and (1.50) yields

(1.51) $\quad 2\alpha'_{pf}Q^{\frac{1}{2}}U = \nabla(\lambda;\Sigma).$

Moreover, the uniqueness of the characteristic vector α^*_{pf} yields

(1.52) $\quad (I_k - (\Sigma^*-\lambda I_k)^+(\Sigma^*-\lambda I_k))U = \alpha^*_{pf}\alpha^{*'}_{pf}\, U$

which equals $\frac{1}{2}\alpha^*_{pf}\nabla(\lambda;\Sigma)$ because of (1.51). In section V.2 it will be shown that the gradient matrix $\nabla(\lambda;\Sigma)$ is given by $\frac{1}{\lambda}(\alpha'_{pf}Q \otimes \alpha'_{pf}Q)$. Substituting (1.51) and (1.52) in (1.49) and replacing $\nabla(\lambda;\Sigma)$ by its outcome, we obtain

(1.53) $\nabla(\alpha_{pf};\Sigma) = -Q^{-1}Q^{\frac{1}{2}}(\Sigma^* - \lambda I_k)^+ Q^{\frac{1}{2}}(\alpha'_{pf}Q \otimes I_k) + \frac{1}{2}Q^{-1}Q^{\frac{1}{2}}\alpha^*_{pf}\frac{1}{\lambda}(\alpha'_{pf}Q \otimes \alpha'_{pf}Q)$,

which is according to some elementary matrix properties equal to the
expression as proposed in (1.44). ◊

Combining (1.42) and (1.44), the limiting distribution of $\hat{\alpha}_{pf}$, (the MME of
α_{pf}) follows directly as a result of the delta method:

(1.54) $\sqrt{N}(\hat{\alpha}_{pf} - \alpha_{pf}) \xrightarrow{D} N(0, [\nabla(\alpha_{pf};\Sigma)]V[\nabla(\alpha_{pf};\Sigma)]')$.

Recall that the convenience of the PSD principle is typically in the form of
the asymptotic covariance matrix of $\hat{\alpha}_{pf}$. The population part is represented by
the gradient matrix of α_{pf}. This matrix is left unchanged by modifications in
the sample design. The stochastic influence of the sample is only reflected in
the asymptotic covariance matrix V of $\sqrt{N}(\text{vec}\hat{\Sigma} - \text{vec}\Sigma)$. An explicit expression
for the asymptotic covariance matrix of $\hat{\alpha}_{pf}$ can for instance be obtained when
the sample consists of N i.i.d. observations, as in this case V is known to
equal the covariance matrix of vec(XX').

Theorem ($\hat{\alpha}_{pf}$ for random i.i.d. sample observations)
Let $\{X_n: n=1,\ldots,N\}$ be a sample of N i.i.d. random observations on X and let
$\hat{\alpha}_{pf}$ be the MME of α_{pf}. If $Q\alpha_{pf}$ is defined as the characteristic vector of Σ
with a squared norm equal to its corresponding unique characteristic root λ in
the metric defined by Q^{-1} (see (1.43)), then   o

(1.55) $\sqrt{N}(\hat{\alpha}_{pf} - \alpha_{pf}) \xrightarrow{D} N(0, (-Q^{-\frac{1}{2}}(\Sigma^* - \lambda I_k)^+ Q^{\frac{1}{2}} + \frac{1}{2}\frac{1}{\lambda}\alpha_{pf}\alpha'_{pf}Q)[E(XX'Q\alpha_{pf}\alpha'_{pf}QXX')]$
$\times (-Q^{\frac{1}{2}}(\Sigma^* - \lambda I_k)^+ Q^{-\frac{1}{2}} + \frac{1}{2}\frac{1}{\lambda}Q\alpha_{pf}\alpha'_{pf}) - \frac{1}{4}\lambda^2\alpha_{pf}\alpha'_{pf})$.

For the proof of this theorem one is referred to the proof of the asymptotic
normality of $\hat{\beta}_{pr}$ in case of an ideal sample (see section III.1), as the matrix
multiplications, yielding their asymptotic covariances, are essentially the
same.
In the preceding theory we have seen that finding a linear subspace that fits

the random vector X as accurately as possible, boils down to deriving the
characteristic vectors of covariance matrix Σ. Defining $Q = I_k$, the analysis
takes place in the well-known Euclidean metric defined by I_k. In the
statistical literature a lot of attention has been given to the large sample
properties of characteristic roots and vectors obtained in this I_k-metric both
in the context of factor analysis and principal components (more about
principal components in the minimum distance setting can be found in section
IV.3 of this study). Although point estimators of the characteristic roots and
vectors mostly are derived in a distribution-free approach (that is, except
for the existence of the covariances no distributional assumptions about the
sample observations are made), the sampling properties of the estimators are
generally obtained for observations from a multivariate normal distribution.
The large sample properties of characteristic roots and vectors under the
normality assumption were for the first time presented by Girshick (1939) and
Anderson (1951). The distribution of the characteristic roots is of importance
for constructing significance tests (see Bartlett (1950, 1951)). As the
characteristic roots of Σ measure the magnitude of variance represented by the
common factors, it is important to know whether they are significantly
different from zero or not. For instance Lawley (1956) has investigated,
without explicitly considering the distribution of the roots, tests for the
smallest roots of Σ. Anderson (1963) presented a review of the results about
large sample distributions of characteristic vectors and tests of hypotheses
for characteristic roots. He also considered a similar study for the
correlation matrix, which is generally used in principal components analysis
as a substitute of the covariance matrix. Those who are interested in more
readings on asymptotic inference about characteristic roots and vectors are
referred to James (1964), Sugiyama (1966), Khatri and Pillai (1969). However,
one should keep in mind that these papers deal with the problem in the context
of the multivariate normal distribution, while in this study only existence of
the fourth-order moments and asymptotic normality of the sample covariance
matrix are assumed.

IV.2 Principal relations versus principal factors

In this section the principal relation (PR) analysis (as discussed in section
III.1) is associated with the principal factor (PF) analysis of the preceding
section. As a result of this comparison not only the relations and differences
between these two techniques will become clear, but also some geometrical
aspects of the minimum distance approach will be elucidated. As this section
is about formerly determined minimization problems, it cannot dispense with
the repetition of some prior defined matrix equations. Quotations of already
used formulas go together with a reference to the equation number, where the
formulas have been used before, and a renumbering for application in this
section. Let us first reconsider the inner minimization problem of the two
techniques. For PR as well as for PF, the squared distance to be minimized is
given by (see (III.1.7) and (1.5))

$$(2.1) \qquad d_Q^2(X, \xi) = (X-\xi)'Q(X-\xi).$$

The equation "$d_Q^2(X, \xi) = $ constant" represents an ellipsoid (in k dimensions)
with the centre at X. In the metric defined by Q it can be considered as a
hypersphere. Distance function (2.1) is minimized with respect to ξ (that is,
the ellipsoid is made as small as possible), subject to the restriction $B'\xi=0$
(in the PR case) or $\xi=A\upsilon$ (in the PF case). Both $B'\xi=0$ and $\xi=A\upsilon$ denote a linear
subspace in R^k and they are of dimension (k-p) (as B is of column rank p) and
dimension q (as A is of column rank q) respectively. Geometrically the
minimization of (2.1) with respect to ξ can be interpreted as the
determination of the non-empty intersection of a linear subspace with the
smallest possible ellipsoid centred at X. The optimal values of these
minimization problems appeared to be (see (III.1.15) and (1.12))

$$(2.2) \qquad \tilde{X} = (I_k - Q^{-1}B(B'Q^{-1}B)^{-1}B')X$$

and

$$(2.3) \qquad \hat{X} = A(A'QA)^{-1}A'QX.$$

The correspondence between these results are emphasized when (2.2) is
rewritten as

(2.4) $\tilde{X} = (I_k - (Q^{-1}B)((Q^{-1}B)'Q(Q^{-1}B))^{-1}(Q^{-1}B)'Q)X.$

In words (2.3) and (2.4) mean:

"$\tilde{X}$ is the Q-orthogonal projection of X on $O_Q(\text{span}(Q^{-1}B))$", and

"$\hat{X}$ is the Q-orthogonal projection of X on span(A)".

Now we come to the outer minimization problems (see (III.1.8) and (1.8))

(2.5) $\min_{B} E(d_Q^2(X,\tilde{X})) \equiv \min_{B} E(X-\tilde{X})'Q(X-\tilde{X})$

and

(2.6) $\min_{A} E(d_Q^2(X,\hat{X})) \equiv \min_{A} E(X-\hat{X})'Q(X-\hat{X}).$

So far, the stochastic character of X has been ignored and A and B are arbitrary matrices. Because of the randomness of X, also $d_Q^2(X,\tilde{X})$ and $d_Q^2(X,\hat{X})$ are random variables. The expected (squared) distance between X and its projection $\hat{X}$ or $\tilde{X}$ is minimized by choosing the optimal projection spaces. That is, in both cases we determine optimal matrices A or B, such that the expected distance between X and the projection space (in which $\tilde{X}$ or $\hat{X}$ are situated) is minimized. In the foregoing theory the optimal values B_{pr} and A_{pf} of B and A were shown to be given by the matrix equations (see (III.1.16) and (1.11))

(2.7) $\Sigma B_{pr} = Q^{-1}B_{pr}\Lambda_{\min,p}$

and

(2.8) $\Sigma Q A_{pf} = A_{pf}\Lambda_{\max,q}.$

Rewriting (2.8) as

(2.9) $\Sigma(QA_{pr}) = Q^{-1}(QA_{pf})\Lambda_{\max,q}$

and observing matrix equations (2.7) and (2.8) in the metric defined by Q^{-1}, the columns of B_{pr} and QA_{pf} can be seen as the characteristic vectors of Σ corresponding to the p smallest and q largest roots respectively. In the special case that $q \le k-p$, the columns of matrix B_{pr} define a set of vectors that is Q^{-1}-orthogonal to the columns of QA_{pf}, or

(2.10) $B'_{pr}Q^{-1}(QA_{pf}) = 0$.

But this is exactly the same as saying that the columns of $Q^{-1}B_{pr}$ are Q-orthogonal to the columns of A_{pf}, yielding the equivalence relation

(2.11) $q \leq k-p$ <=> $span(A_{pf}) \subseteq O_Q(span(Q^{-1}B_{pr}))$.

In a similar way we find

(2.12) $q \geq k-p$ <=> $span(A_{pf}) \supseteq O_Q(span(Q^{-1}B_{pr}))$

and consequently the equality $span(A_{pf}) = O_Q(span(Q^{-1}B_{pr}))$ holds if and only if $q = k-p$. Combining (2.11) and (2.12) with the verbal descriptions of the projections of X defined in (2.3) and (2.4) it follows that for $q = k-p$, $\tilde{X}$ and $\hat{X}$ are identical projections of X on a q-dimensional linear subspace. Hence for $q = k-p$ the two mathematical expressions $B'_{pr}\tilde{X} = 0$ and $\hat{X} = A_{pf}U$ are two different equations, that describe one and the same linear relationship.

In section III.2 the distance matrix Q was studied more closely. It was argued that the form of Q defines the projection direction. Special attention was given to the case that Q is a diagonal matrix where one (or in general k-m) of its diagonal elements tend to infinity. It appeared that an infinite diagonal element caused the corresponding dimension of X to be unchanged under the projection in the metric defined by Q. In other words, when the ith diagonal element of Q tends to infinity, then the ith component of the optimal value $\tilde{X}$, obtained by minimization of (2.1), equals X_i, the ith component of X. X_i was called an exogenous variable of the linear structural model. In the more general case of a non-diagonal Q, the infinity of a characteristic root excludes a dimension from the projection direction. We already noted that the objective function of the inner minimization problem given in (2.1) as a function of ξ, defines an ellipsoid in the simple Euclidean metric, centred at X. The projection direction can be denoted by a vector from the centre of the ellipsoid (X) to its surface. Clearly the ellipsoid will be degenerated when one of the characteristic roots of Q equals zero or tends to infinity. Let us for simplicity observe a diagonal metric matrix in 3 dimensions, defined by

$$(2.13) \quad Q_* = \begin{bmatrix} 1 & 0 & 0 \\ 0 & 1 & 0 \\ 0 & 0 & q \end{bmatrix}$$

A graphical representation of ellipsoid

$$(2.14) \quad (X-\xi)'Q_*(X-\xi) = d^2$$

subject to the assumption that $E(X)=0$, is given in figure (IV.2.1).

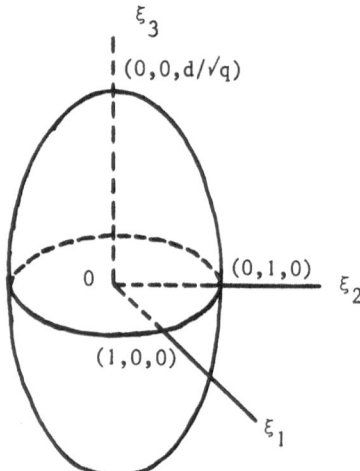

Figure IV.2.1 The 3-dimensional ellipsoid $\xi'Q_*\xi = d^2$ with $Q = \text{diag}(1,1,q)$

The principal axes of this ellipsoid correspond with the coordinate axes, as Q is a diagonal matrix. The intersection of the ellipsoid is circular in the plane defined by $\xi_3 = 0$, as the first two diagonal elements of Q are the same. The length of the third principal axis is proportional to the square root of q^{-1}, hence the height of the ellipsoid depends on the magnitude of q. Clearly the ellipsoid degenerates into a cylinder when $q \to 0$ and only a circle in the plane generated by the ξ_1- and ξ_2-axis remains when $q \to \infty$. Hence the assumption of an exogenous variable in a linear structural model makes the set of points with equal Q-distance to X degenerate into a circle in the (ξ_1, ξ_2)-plane. In general, for a non-diagonal (k×k) distance matrix Q, the principal axes of the ellipsoid formed by the equi-distance points, are determined by

the characteristic vectors and roots of Q. It can be proven that for a Q with
all different characteristic roots $\lambda_1 > \ldots > \lambda_k$, the directions of the principal
axes are given by the characteristic vectors and the length of the ith
principal axis equals $d/\sqrt{\lambda_i}$ (see for instance Mardia, Kent, Bibby (1979, p.
484)). As the feasible points of the inner minimization problem are restricted
to a (k-p)- respectively q-dimensional linear subspace, not all the points
with minimum Q-distance to X are allowed as optimal value of ξ. This
restriction is graphically represented by the intersection of the ellipsoid by
a (k-p)- respectively q-dimensional hyperplane through the origin. In figure
(IV.2.2) an example is given for k=3 and p=1 (or, equivalently q=2).

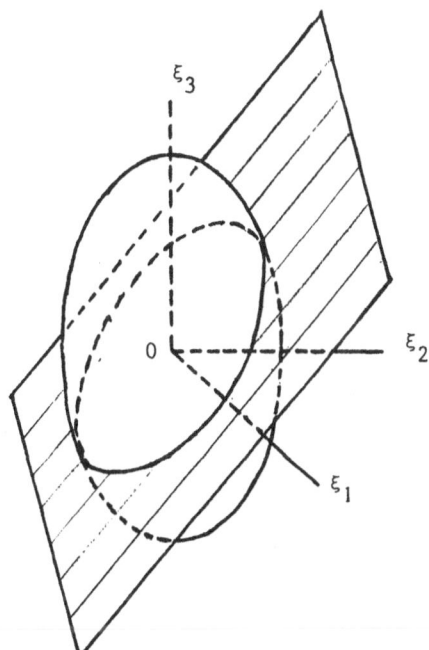

Figure IV.2.2 Intersection of the 3-dimensional ellipsoid $\xi'Q\xi = d^2$, where
Q = diag(q_1, q_2, q_3), with a 2-dimensional plane $\beta'\xi = 0$.

The resulting set of feasible points form a 2-dimensional ellips in a 3-
dimensional space. In general for $k \geq 3$ and $1 \leq q < k$, the set of equi-
distance points subject to the q-dimensional linear restriction defines a q-
dimensional ellipsoid in a k-dimensional space. In the outer minimization
problem, the optimal directions of the restrictive hyperplane are determined.

Similar to the assumption of exogenous variables in the PR analysis, yielding
the so-called simultaneous equations system and linear regression analysis, we
can define exogenous variables in the PF analysis, by setting corresponding
diagonal elements of Q equal to infinity. Indicating the endogenous part of X
by (m×1) vector Y and the exogenous part by ((k-m)×1) vector Z, X is given by
(see also (III.3.2))

$$(2.15) \qquad X = \begin{bmatrix} Y \\ Z \end{bmatrix}$$

and the metric defining matrix is given by (see als (III.3.6))

$$(2.16) \qquad Q = \begin{bmatrix} Q_{yy} & 0 \\ \hline 0 & \begin{matrix} \infty & \\ & \ddots \\ & & \infty \end{matrix} \end{bmatrix} .$$

Defining analogous partitionings for the instrumental vector ξ and the
parameter matrix A

$$(2.17) \qquad \xi = \begin{bmatrix} \eta \\ \zeta \end{bmatrix} , \quad A = \begin{bmatrix} A_y \\ \hline A_z \end{bmatrix} ,$$

the PF composite minimization problem (1.9) can be rewritten as

$$(2.18) \qquad \min_{\substack{A \\ Z = A_z \upsilon}} E[\min_{\upsilon} \ (Y - A_y \upsilon)' Q_{yy} (Y - A_y \upsilon)].$$

Hence the assumption of unbounded diagonal elements in Q yields a minimization
problem in the endogenous part Y of X with additional restriction $Z = A_z \upsilon$.
Because of the projection of the Y-vector on a best fitting hyperplane and the
restriction $Z = A_z \upsilon$, some assumptions concerning the various dimensions have
to be made. As the projection of Y implies a dimension reduction, it is
necessary to presuppose $q \leq m$. Furthermore, in order to let the system of

equations $Z = A_z \upsilon$ be compatible (that is, there exists at least one solution for A_z, when υ is replaced by its optimal value), it is assumed that the number of Z-components is less than or equal to the number of components of υ. Hence $(k-m) \leq q$. The two dimension assumptions together are given by

$$(2.19) \quad k-m \leq q \leq m.$$

Optimal values for υ, A and η ($= A_y \upsilon$) are derived by applying the same differentiation techniques as those used in section 1. However, already the solution of the inner minimization of problem (2.18) yields unpleasant and outstretched formulas. For instance the optimal U of υ is given by

$$(2.20) \quad U = (A'_y Q_{yy} A_y)^{-1} A'_y Q_{yy} Y$$

$$- (A'_y Q_{yy} A_y)^{-1} A'_z (A_z (A'_y Q_{yy} A_y)^{-1} A'_z)^{-1} (A_z (A'_y Q_{yy} A_y)^{-1} A'_y Q_{yy} Y - Z).$$

(Note, that assuming the matrices A_y and A_z to be of full column rank, the inverses in (2.20) exist because of the dimension assumptions in (2.19)). Because of the non-interpretability of this result and the laborious resulting objective function of the outer minimization problem, the case of exogenous variables in PF analysis will not be worked out in this study.

In the various sections of chapter III, PR analysis has been worked out for some special a priori choices of matrix Q. As said before, this will not be done in this chapter for the PF case; in the section to come some attention will be given to a multivariate statistical technique, that is closely related to PF analysis. This technique is in the statistical literature known under the name principal components analysis and it is, like factor analysis, an attempt to describe a random vector by a linear combination of vectors of lower dimension. It will be shown, that the minimization problem, corresponding with principal components analysis, is closely related with the minimum distance problem of PF analysis, as described in section 1.

IV.3 Principal components analysis

In section 1 we considered a method called principal factor (PF) analysis, in order to obtain a linear subspace in R^k, that fits the distribution of the random vector X as accurately as possible. The optimal subspace was found to be spanned by the characteristic vectors that correspond with the q maximal characteristic roots of covariance matrix Σ. A random (q×1) vector U of latent variables was defined, such that $A_{pf}U$ equals the projection of X onto the best fitting hyperplane. Hence in PF analysis the vector $A_{pf}U$ represents X in less dimensions, such that it minimizes the expected squared distance (in the Euclidean metric defined by Q) to X. The purpose of principal components (PC) analysis is also to reduce the dimensionality of X, however, now optimality properties are defined in terms of variances. PC analysis considers the transformation of X to a new set of unobservable variables, which are mutually uncorrelated and that have maximum possible variance. In this way the variability of X is characterized as good as possible by a vector that consists of less components. This technique originates from Hotelling (1933), who generalized the idea of reducing the dimensionality of multivariate data encountered in a sample (see also Pearson (1901)) to dimension reduction of a random vector.

Formally the linear combinations of X are given by A'X, where A is a (k×q) parameter matrix, with q denoting the required reduced dimensionality (hence q < k). Denoting the columns of A by the (k×1) vectors $\alpha_1, \ldots, \alpha_q$, that is

(3.1) $A = [\alpha_1, \ldots, \alpha_q]$ with α_i a (k×1) vector for i=1,...,q,

the optimal value of A is obtained as the solution of q mathematical programming problems, given for i=1,...,q by

(3.2) $\max_{\alpha_i} \alpha_i' \Sigma \alpha_i$

subject to $\alpha_i' \Sigma \alpha_j = 0$ for j=1,...,i-1

$\alpha_i' \alpha_i = 1.$

In words, the ith maximization problem in (3.2) stands for the determination of the linear compound $\alpha_i'X$ with maximum variance, subject to the i-1 restrictions that the resulting random variable is uncorrelated with the predetermined linear combinations $\alpha_1'X, \ldots, \alpha_{i-1}'X$. The last restriction $\alpha_i'\alpha_i = 1$ simply requires that the vector α_i is of length 1 in the simple Euclidean metric. This (unit-)normalization is needed as the objective function can be made arbitrary large by multiplying a feasible solution for α_i by a constant and consequently a scaling is needed to assure the uniqueness of the optimal vector. For thourough treatment and detailed accounts on maximization problem (3.2) one is referred to statistical textbooks like Anderson (1958, chapter 11), Morrison (1978, chapter 8) and Muirhead (1982, chapter 9). They prove that the optimal value A_{pc}, say, of matrix A in the PC problem (3.2) is given by

$$(3.3) \qquad \Sigma A_{pc} = A_{pc} \Lambda_{max,q} \quad \text{and} \quad A_{pc}' A_{pc} = I_q \,,$$

with Σ the covariance matrix of X and $\Lambda_{max,q}$ the diagonal matrix with on its diagonal the q largest characteristic roots of Σ (in the simple Euclidean metric). Hence the columns of A_{pc} equal the normalized characteristic vectors of Σ. From (3.3) it follows directly that the covariance matrix of the resulting q optimal linear transformations of X is given by

$$(3.4) \qquad A_{pc}' \Sigma A_{pc} = \Lambda_{max,q}.$$

Hence the variances of the components of $A_{pc}'X$ are given by the maximum characteristic roots of Σ. The linear combination $\alpha_{pc,i}'X$ is called the ith principal component and the corresponding characteristic root λ_i reflects the amount of the total variance $tr(\Sigma)$ of the components of X, represented by the ith principal component. Note that all the variance in the components of X is accounted for by the principal components if q=k, for then

$$(3.5) \qquad tr(\Lambda_k) = tr(A_{pc}' \Sigma A_{pc}) = tr(\Sigma A_{pc} A_{pc}') = tr(\Sigma).$$

The theory of PC is related to the theory of CC (see section III.6) in the sense that they both maximize (co)variances of linear combinations of X, subject to some restrictions concerning the independency of the resulting various linear compounds. In the former, the problem is to define a number of mutually uncorrelated linear combinations of X, exhibiting maximal variance, while in CC analysis the problem is to define two sets of variables, obtained as linear combinations of Y and Z with $X = (Y',Z')'$, such that within each set the variables are uncorrelated while the corresponding elements of the two sets exhibit maximum correlation. In section III.6 it has been argued that the covariance maximizing CC problem yields the same optimal solutions as the distance minimizing principal relations (PR) problem, when a specific metric matrix Q is chosen. In a similar way one may show that the variance maximizing PC problem yields the same optimal solution as the distance minimizing PF problem when Q is defined to equal the unit matrix I_k and with unit normalization for the optimal solution of A. Merely the fact that both PC analysis and PF analysis are fundamentally concerned with the largest characteristic roots and vectors of Σ, makes one think of an association between the two optimization problems. The equivalence between the two optimization problems can analytically be reasoned as follows. The outer minimization of PF with $Q = I_k$ and the normalization restriction (see also (1.15)) is given by

$$(3.6) \qquad \min_{A} \ tr(\Sigma - A(A'A)^{-1}A'\Sigma)$$

$$\text{subject to } A'A = I_q.$$

Substitution of the restriction in the objective function, neglecting its first term and using trace operator properties, (3.6) may be rewritten as

$$(3.7) \qquad \max_{A} \ tr(A'\Sigma A)$$

$$\text{subject to } A'A = I_q.$$

This is not yet exactly the PC problem defined in (3.2), as the restrictions on α_i and α_j, for $i \neq j$, are not the same. However, from section 1 it is known that the optimal solution for A, here indicated by a subscript "pc", is given by (see also (1.11) with $Q = I_k$)

$$(3.8) \qquad \Sigma A_{pc} = A_{pc}\Lambda_{max,q} \quad \text{and} \quad A'_{pc}A_{pc} = I_q$$

which yields

(3.9) $A'_{pc} \Sigma A_{pc} = \Lambda_{max,q}$.

Hence the restriction $\alpha'_{pc,i} \alpha_{pc,j} = 0$ (for $i,j=1,\ldots,q$ and $i \neq j$) is implied
by the optimal value of the PF problem in the I_k-metric. Conversely, it can
easily be seen that restrictions $\alpha'_{pc,i} \alpha_{pc,j} = 0$ (for $i,j=1,\ldots,q$ and $i \neq j$) in
the PF problem given by (3.7) are implied by the optimal solution (3.3) of the
PC problem. As a consequence we may conclude that the PF problem with $Q = I_k$
and the PC problem, both with unit normalization of the columns of parameter
matrix A, yield the same solution. A restriction for the expected distance
minimization is found to be a property of the variance maximizing solution and
inversely the restriction for the variance maximization appears to be a
property of the distance minimizing solution. In this way we have found two
optimality properties of the principal components. Various other ways of
interpreting the principal components have for instance been presented by Rao
(1964), Darroch (1965) and Okamoto and Kanazawa (1968). A review on the
optimal properties of principal components has been given by Okamoto (1969).
As a result the principal components can be obtained as a solution of a
minimum distance problem. Reformulating the PF problem (see also (1.9)) as

(3.10) $\min_A E[\min_\upsilon (X-A\upsilon)'(X-A\upsilon)]$.

The principal components are given by the following theorem.

Theorem (the PC matrix)
Let the mathematical programming problem be defined by (3.10) with

    X a (k×1) random vector
    υ a (q×1) auxiliary vector
    A a (k×q) parameter matrix such that $A'A = I_q$.

The optimal value U of υ is then given by

(3.11) $U = A'X$

and the optimal value A_{pc} of A is defined by

(3.12) $\Sigma A_{pc} = A_{pc} \Lambda_{max,q}$, $A'_{pc} A_{pc} = I_q$

with Σ the covariance matrix of X and $\Lambda_{max,q}$ the diagonal matrix with on its diagonal the q largest characteristic roots of Σ in the metric defined by I_k.

In section 1 we observed a corresponding minimization problem in an arbitrary Euclidean metric defined bij Q. Hence the proof of this theorem follows directly from the derivation of A_{pf} in section 1, by setting the matrix Q equal to I_k. Furthermore, it is to be noted, that A_{pc} is normelized such that $A'_{pc} A_{pc} = I_q$, while in PF analysis $A'_{pf} Q A_{pf} = \Lambda_{max,q}$ is the most natural normalization.

Observing (3.11) it follows that when the matrix A is replaced by its optimal value A_{pc}, the vector U denotes the principal components of X, that are uncorrelated and have variances that equal the q maximum characteristic roots of Σ. From this expression the main difference between the interpretations of PF and PC becomes clear. In PF analysis the emphasis is on a transformation from the unobservable factors U to the observed variables X ($\hat{X} = AU$, see (1.7)), while in PC analysis the emphasis is on a transformation from the observable vector X to the unobservables in U (U = A'X, see (3.11)). The development of PC analysis as a technique that determines the linear combinations of X, that represent the main part of the variance in X, gives rise to a second difference between PC and PF analysis. (The following remark can be found in most textbooks that deal with both PC and factor analysis). So far, no assumptions have been made about the units in which the components of X are measured. Therefore, it is difficult to attach a meaning to the concept of partitioning the total variation into the contributions due to each principal component. Measuring a component of X in smaller units will increase its relative contribution to the total variation, if the measurement of the other components is kept unchanged. Therefore the influence of the components on the total variation of X depends on the measurement units. In order to avoid this problem, it is common practice to standardize the vector X, before carrying out the analyses. Σ is now the matrix of correlation coefficients of X, that has in section II.2 been denoted by R. The relation between ρ_{ij} (the (i,j)[th] element of R) and σ_{ij} (the (i,j)[th] element of Σ) is given by

(3.13) $\rho_{ij} = \dfrac{\sigma_{ij}}{\sqrt{\sigma_{ii}}\,\sqrt{\sigma_{jj}}}$ $(i,j=1,\ldots,k)$.

The characteristic vectors are not invariant under rescaling of the variables, hence the solutions for Σ and R differ in a non-trivial way. Except for the replacement of Σ by R, all arguments for PC analysis remain the same.

Now we come to the derivation of the limiting distribution of the method of moments estimator of A_{pc}. Similar to the method applied in the preceding sections, this will be done by assuming the existence of an asymptotically normally distributed estimator $\hat{\Sigma}$ of Σ and applying the delta method. In case of standardized observations of X, one should not proceed from $\hat{\Sigma}$, but from an asymptotically normally distributed estimator of correlation matrix R, say $\hat{R}$. Replacing in (3.13) the (co)variances σ_{ij}, σ_{ii} and σ_{jj} by their sample counterparts, we obtain

(3.14) $\hat{\rho}_{ij} = \dfrac{\hat{\sigma}_{ij}}{\sqrt{\hat{\sigma}_{ii}}\,\sqrt{\hat{\sigma}_{jj}}}$ $(i,j=1,\ldots,k)$,

and the $(k \times k)$ sample correlation matrix $\hat{R}$ is defined by

(3.15) $\hat{R} = [\hat{\rho}_{ij}]$.

In section II.2 (see (II.2.22)) it has been shown that the asymptotic normality of $\text{vec}\hat{R}$ follows from the asymptotic normality of $\text{vec}\hat{\Sigma}$ and the components of the covariance of the limiting distribution, in case of a sample consisting of i.i.d. observations, are given in (II.2.21). In order to keep the notation consistent with the foregoing sections, we shall proceed by using the covariance matrix Σ, rather than the correlation matrix R. However, from the aforementioned theorem that deals with the asymptotic normality of $\text{vec}\hat{R}$, corresponding results for standardized observations follow directly. The MME $\hat{A}_{pc}$ of A_{pc} is obtained by replacing in (3.12) the covariance matrix Σ by its estimator $\hat{\Sigma}$, yielding

(3.16) $\hat{\Sigma}\hat{A}_{pc} = \hat{A}_{pc}\hat{\Lambda}_{\max,q}$ with $\hat{A}'_{pc}\hat{A}_{pc} = I_q$.

The asymptotic normality of $\text{vec}\hat{A}_{pc}$ is obtained by means of the delta method. That is, when the asymptotic covariance matrix of $\sqrt{N}(\text{vec}\hat{\Sigma} - \text{vec}\Sigma)$ is indicated by V, then the limiting distribution of $\text{vec}\hat{A}_{pc}$ is given by

$$(3.17) \qquad \sqrt{N}(\text{vec}\hat{A}_{pc} - \text{vec}A_{pc}) \overset{D}{\to} \mathbf{N}(0, \ [\nabla(A_{pc};\Sigma)]V[\nabla(A_{pc};\Sigma)]')$$

where $\nabla(A_{pc};\Sigma)$ denotes the $(kq \times k^2)$ matrix of partial derivatives of $\text{vec}A_{pc}$ with respect to $(\text{vec}\Sigma)$. In the following theorem the elements of the required gradient matrix are given for one column of A_{pc}.

<u>Theorem</u> (the derivatives of the PC matrix)

Let λ and α_{pc} be a characteristic root and the corresponding unique, normalized characteristic vector of Σ in the metric defined by I_k, that is

$$(3.18) \qquad \Sigma\alpha_{pc} = \alpha_{pc}\lambda \quad \text{and} \quad \alpha'_{pc}\alpha_{pc} = 1.$$

Then α_{pc} may be considered as a differentiable function of Σ with

$$(3.19) \qquad \nabla(\alpha_{pc};\Sigma) = -(\alpha'_{pc} \otimes (\Sigma - \lambda I_k)^+)$$

where $(\Sigma - \lambda I_k)^+$ is the Moore-Penrose inverse of $(\Sigma - \lambda I_k)$.

<u>Proof</u>

In section III.1 the partial derivatives of the principal relations vector β_{pr} were analysed, where β_{pr} denotes the normalized characteristic vector of Σ in the Q^{-1}-metric, corresponding with the smallest characteristic roots. Note that both α_{pc} and β_{pr} are characteristic vectors of Σ. The first in the Euclidean metric defined by I_k and the latter in the more general Q^{-1}-metric. The fact that β_{pr} corresponds with the smallest characteristic roots and α_{pc} corresponds with the largest characteristic roots of Σ is of no influence on the form of their partial derivatives. Hence the matrix of partial derivatives denoted in (3.19) are simply obtained by substituting the unit matrix I_k for matrix $Q^{\frac{1}{2}}$ and replacing β_{pr} by α_{pc} in the matrix $-(\beta'_{pr} \otimes Q^{\frac{1}{2}}(\Sigma^* - \lambda I_k)^+ Q^{\frac{1}{2}})$, which was shown to be the matrix of partial derivatives in the PR case (see (III.1.37)). $\diamond$

In the preceding sections special attention has been given to the case that
the sample consist of N i.i.d. observations. This so-called ideal sample
situation is an example of a possibility to replace the asymptotic covariance
matrix V by some explicit expressions, using fourth-order central moments of
X. Because of the analogy between the partial derivatives of the PC and the PR
parameters, the asymptotic covariance matrix of $\hat{\alpha}_{pc}$ is easily derived from the
ideal sample results for $\hat{\beta}_{pr}$ given in (III.1.54). Replacing in (III.1.54)
$\hat{\beta}_{pr}$ by $\hat{\alpha}_{pc}$ and Q by I_k the limiting distribution of $\hat{\alpha}_{pc}$ for N i.i.d.
observations is obtained. However, we have seen that, in order to provide a
meaningful interpretation for PC analysis, one should work with standardized
observations. That is, the middle matrix V has to be replaced by the
covariance matrix of the cross products of the standardized random vector X,
yielding asymptotic distribution properties of sample correlation matrix
$\hat{R}$. In section II.2 it has been shown that $vec\hat{R}$ is asymptotically normally
distributed and the elements of the covariance matrix W of the limiting
distribution are given in (II.2.21). As it is difficult to present the results
in matrix notation it seems wise to preserve the pre- and post-multiplication
of W by the derivative of α_{pc} for the computer.

So far, this study has restricted the statistical inference to asymptotic
properties of the characteristic vectors of covariance and correlation
matrices. The large sample properties of the characteristic roots have not
been dealt with yet. In the next chapter it will be observed, that the
characteristic roots play an important role in defining a goodness-of-fit
measure for the linear structural model. It has already been noted that for
instance in PC analysis the largest characteristic roots reflect the amount of
the total variance $tr(\Sigma)$ of the components of X represented by the
corresponding principal components. Therefore, the characteristic roots form
an indication of how much the principal components contribute to the
explanation of the variation in X. It is for this reason that the derivation
of the large sample properties of the characteristic roots will be postponed
until the intuitive minimum distance interpretation of these covariance matrix
characteristics has been given.

V. Goodness-of-Fit measures

V.1 Coefficients of multiple correlation and angles between random vectors

In the preceding chapters statistical, structural linear models have been discussed in a minimum distance setting. The linear relations between the components of the random vector X are described by the model parameters, which are in the Principal Relations (PR) case represented by a (k×p) matrix B and in the Principal Factors (PF) case by means of a (k×q) matrix A. Application of the method of moments technique yields consistent estimators of B and A and using the delta method, their limiting distributions can be obtained. The asymptotic variances of the parameter estimators give some information on the spread of the values of the estimates about the population value. That is, the variance indicates how well the population parameter is estimated by this specific sample. But it does not say anything about the credibility of the linear specification itself.

The aim of this chapter is to define a coefficient that indicates the credibility of the hypothesized linear description of the relations in X. The estimator of this coefficient measures the goodness-of-fit of the model on a scale of 0 to 1. A familiar example of a goodness-of-fit measure is the multiple correlation coefficient, that is derived in the context of a functional linear regression model. In this section the multiple correlation coefficient is considered more closely. It is indicated in which way this measure is a generalization of the correlation coefficient between two random variables. The multiple correlation coefficient can be generalized in a similar way to goodness-of-fit measures for sets of linear relations. The problem of deriving goodness-of-fit measures for simultaneous equations was already discussed by Hooper (1959, 1962) and Glahn (1969) in the functional linear model context. McElroy (1977) presented a measure of goodness-of-fit for Zellner's seemingly unrelated regressions and compared his results to Hooper's and Glahn's correlation coefficients. The treatise of the goodness-of-fit measures starts by a short review of the aforementioned literature (see also Dhrymes, 1980 section 5.4) and afterwards a goodness-of-fit measure will be introduced for the structural models that have been discussed in the chapters III and IV. In the remaining three sections of this chapter, the goodness-of-fit measure will be discussed subject to different assumptions for the distance defining matrix Q and parameter matrices B and A, yielding goodness-of-fit measures for various types of PR and PF analysis.

In the literature there is confusion about what is a good measure of fit, except in the case of OLS-regression. This may be the reason that, except for OLS results, there is a reluctancy to present goodness-of-fit measures for empirical results. Let us start by observing the well-known multiple correlation coefficient, usually denoted by R^2, of the functional linear regression model. Defining y_n to be the dependent variable and z_n to be the explanatory vector respectively, the linear regression model is given by

$$(1.1) \qquad y_n = \gamma + b'z_n + \varepsilon_n \qquad\qquad (n=1,\ldots,N)$$

where the residuals $\varepsilon_1,\ldots,\varepsilon_N$ have zero mean, identical variances and zero covariances. The variable y_n and vector z_n can be considered as components of the vector $x_n = (y_n, z_n')'$. Let the vector of sample means be denoted by

$$(1.2) \qquad \bar{x}_n = \begin{bmatrix} \bar{y}_n \\ \bar{z}_n \end{bmatrix} = \frac{1}{N} \sum_{n=1}^{N} x_n \qquad\qquad (k\times 1)$$

and the matrix of sample cross-products by

$$(1.3) \qquad S = \begin{bmatrix} S_{yy} & S'_{zy} \\ S_{zy} & S_{zz} \end{bmatrix} = \frac{1}{N} \sum_{n=1}^{N} (x_n - \bar{x})(x_n - \bar{x})' \qquad (k\times k).$$

Then the OLS estimator of regression vector b is given by

$$(1.4) \qquad \hat{b} = S_{zz}^{-1} S_{zy}$$

and the multiple correlation coefficient is defined as

$$(1.5) \qquad R^2 = 1 - \frac{\frac{1}{N} \sum_{n=1}^{N} e_n^2}{S_{yy}} \qquad\qquad \text{with } e_n = y_n - \hat{b}'z_n.$$

S_{yy} is the total sum of squares and $\frac{1}{N} \sum_{n=1}^{N} e_n^2$ is the sum of squared errors, representing the portion of the variation in $y_1,\ldots,y_N$ about $\bar{y}$ that is not explained by the linear regression model. Consequently, the R^2 is frequently referred to as the "coefficient of determination" or the "degree of explanation", as it expresses the sample proportion of $y_1,...,y_N$ that is explained by the linear model. More about the use of R^2 in the search for correct model specifications context can, for instance, be found in Cramer (1964) and Koerts and Abrahamse (1970).

Replacing in (1.5) e_n by $y_n - \hat{b}'z_n$ it can be shown that

$$(1.6) \qquad R^2 = \frac{\hat{b}'S_{zz}\hat{b}}{S_{yy}} = \frac{S'_{zy}S_{zz}^{-1}S_{zy}}{S_{yy}}.$$

Before we interpret this notation of R^2, let us first recall the definition of correlation coefficient ρ_{yz} of two random variables Y and Z. When the variances and covariances of Y and Z are denoted by σ_{yy}, σ_{zz} and σ_{zy}, then

$$(1.7) \qquad \rho_{yz} = \frac{\sigma_{zy}}{\sqrt{\sigma_{yy}}\sqrt{\sigma_{zz}}} \quad \text{or} \quad \rho_{yz}^2 = \frac{\sigma_{zy}^2}{\sigma_{yy}\sigma_{zz}}.$$

Replacing in (1.7) the (co)variances by their sample analogues s_{yy}, s_{zz} and s_{zy}, say, then the sample counterpart r_{yz}^2 of ρ_{yz}^2 is obtained, yielding

$$(1.8) \qquad r_{yz}^2 = \frac{s_{zy}^2}{s_{yy}s_{zz}} = \frac{s_{zy}s_{zz}^{-1}s_{zy}}{s_{yy}}.$$

Note the correspondence between R^2 in (1.6) and r_{yz}^2 in (1.8). R^2 can be observed as a generalization of the sample correlation coefficient r_{yz}^2, denoting a (multiple) correlation between scalar variables $y_1,...,y_n$ and vector variables $z_1,...,z_n$. The population analogue of (1.6) is given by

$$(1.9) \qquad \rho_{yz}^2 = \frac{\Sigma'_{zy}\Sigma_{zz}^{-1}\Sigma_{zy}}{\Sigma_{yy}}$$

where Σ_{yy}, Σ_{zz} and Σ_{zy} stand for blocks in the covariance matrix Σ, denoting

the (co)variances between random variable Y and random vector Z. Hence ρ^2_{yz} is
a measure of linear association between Y and Z, that lies in the interval
$[0,1]$.

In a similar way, the correlation coefficient can be extended to a parameter
that measures correlation between two vectors. Let us therefore define X to be
a random vector partitioned into two sub-vectors Y and Z, consisting of m and
(k-m) variables respectively, and let the covariance matrix of X be similarly
partitioned as

$$(1.10) \qquad \Sigma = \left[\begin{array}{c:c} \Sigma_{yy} & \Sigma'_{zy} \\ \hdashline \Sigma_{zy} & \Sigma_{zz} \end{array}\right].$$

The coefficient of correlation between two vectors is then defined by (see
also Dhrymes (1980, p. 250))

$$(1.11) \qquad \rho^2_{yz} = \frac{\left|\Sigma'_{zy}\Sigma^{-1}_{zz}\Sigma_{zy}\right|}{\left|\Sigma_{yy}\right|} = \left|\Sigma^{-1}_{yy}(\Sigma'_{zy}\Sigma^{-1}_{zz}\Sigma_{zy})\right|.$$

($\left|\Sigma_{yy}\right|$ is also known as the generalized variance of Y, see also Wilks (1962,
chapter 18)). Let us first assume that the number of variables in Y do not
exceed the number of variables in Z (m $\leq$ k-m). Obviously ρ^2_{yz} equals 0 if Y and
Z are uncorrelated and ρ^2_{yz} equals 1 if there exists an exact linear relation
between Y and Z. That the value of ρ^2_{yz} always lies in the interval $[0,1]$ can
be seen as follows. The determinant of $\Sigma^{-1}_{yy}(\Sigma'_{zy}\Sigma^{-1}_{zz}\Sigma_{zy})$ equals the product of
its characteristic roots, or, in other words

$$(1.12) \qquad \rho^2_{yz} = \prod_{i=1}^{m} \phi^2_i$$

with ϕ_i the ith characteristic root of $\Sigma'_{zy}\Sigma^{-1}_{zz}\Sigma_{zy}$ in the Euclidean metric
defined by Σ_{yy}. From section III.6 (see (6.11) and (6.16)) it is known that ϕ_i
is the ith canonical correlation. This means that $-1 \leq \phi_i \leq 1$ and from (1.12)
it then follows that $0 \leq \rho^2_{yz} \leq 1$. Hooper (1962) used this parameter as a
goodness-of-fit measure for the reduced form of simultaneous equations
systems. He also pointed out the disadvantage of this correlation coefficient,

viz. if $m > k-m$, then $\Sigma'_{zy}\Sigma^{-1}_{zz}\Sigma_{zy}$ is a singular matrix and $\rho^2_{yz} = 0$, irrespective of the value of Σ_{zy}. In order to avoid this problem he suggested to take the average of $\phi^2_1,\ldots,\phi^2_m$, rather than their product as a goodness-of-fit measure of the reduced form linear model, yielding the so-called trace correlation that is defined by

$$(1.13) \qquad \bar{\rho}^{-2}_{yz} = \frac{1}{m} \sum_{i=1}^{m} \phi^2_i = \frac{1}{m} \, tr(\Sigma^{-1}_{yy}(\Sigma'_{zy}\Sigma^{-1}_{zz}\Sigma_{zy})).$$

The trace correlation coefficient is 0 if and only if $\phi_1 = \ldots = \phi_m = 0$, which will be so only if $\Sigma_{zy} = 0$.

A coefficient that can be seen as the opposite of the aforementioned correlation coefficient is referred to as the coefficient of alienation (see also Dhrymes (1980, p. 247)). This coefficient of alienation between two vectors Y and Z is given by

$$(1.14) \qquad \tau^2_{yz} = \frac{|\Sigma|}{|\Sigma_{yy}| \, |\Sigma_{zz}|}.$$

Note that when Y and Z are uncorrelated (hence $\Sigma_{zy} = 0$), then $|\Sigma| = |\Sigma_{yy}| \, |\Sigma_{zz}|$ and as a result τ^2_{yz} equals 1. When an exact linear relation holds between Y and Z, then $|\Sigma| = 0$ and also $\tau^2_{yz} = 0$. Formally we may say

$$(1.15) \qquad (\tau^2_{yz} = 0 \iff \rho^2_{yz} = 1) \quad \text{and} \quad (\tau^2_{yz} = 1 \iff \rho^2_{yz} = 0).$$

Using the relation

$$(1.16) \qquad |\Sigma| = |\Sigma_{zz}| \, |\Sigma_{yy} - \Sigma'_{zy}\Sigma^{-1}_{zz}\Sigma_{zy}|,$$

the coefficient of alienation can be rewritten as

$$(1.17) \qquad \tau^2_{yz} = \frac{|\Sigma_{yy} - \Sigma'_{zy}\Sigma^{-1}_{zz}\Sigma_{zy}|}{|\Sigma_{yy}|} = |I_m - \Sigma^{-1}_{yy}(\Sigma'_{zy}\Sigma^{-1}_{zz}\Sigma_{zy})|.$$

Defining ϕ_i as the ith canonical correlation of Y and Z, (1.17) yields

(1.18) $\tau_{yz}^2 = \prod_{i=1}^{m} (1-\phi_i^2).$

Similar to the trace correlation, one can define the trace analogue of the coefficient of alienation as

(1.19) $\bar{\tau}_{yz}^2 = \frac{1}{m} \sum_{i=1}^{m} (1-\phi_i^2).$

In the context of a set of regression equations the coefficients of alienation are used to measure the relative residual variability, remaining after the influence of the explanatory variables on the jointly dependent variables have been removed. Therefore it seems reasonable that the coefficient of correlation and the coefficient of alienation sum to 1. Although this property holds for the trace analogues of the coefficients, that is

(1.20) $\bar{\rho}_{yz}^2 + \bar{\tau}_{yz}^2 = 1,$

this property generally does not hold for ρ_{yz}^2 and τ_{yz}^2, unless all the canonical correlations equal 0 or equal 1.

Let us return to the minimum distance approach of the structural linear statistical model. For the model in its most general form (indicated by PR and PF analsysis), no distinction is made between endogenous and exogenous, or dependent and explanatory variables. The model merely describes a best fitting hyperplane for the random vector X by means of the parameter matrices B_{pr} or A_{pf}. As the goodness-of-fit of the model is in the distance between X and $\tilde{X}$ (or $\hat{X}$), it seems reasonable to take the divergence of X from its projection as an indication of how much of the linear association $B'\tilde{X} = 0$ or $\hat{X} = AU$, that holds exactly for $\tilde{X}$ or $\hat{X}$, also holds for X itself. The minimum distance concept will be used to define a measure of goodness-of-fit between X and $\tilde{X}$. (For notational convenience, the introduction of the goodness-of-fit measure is only conceptualized for $\tilde{X}$. The arguments also hold for $\hat{X}$. The corresponding results for the PF case will be given at the end of this section). The expected squared distance between X and its projection $\tilde{X}$ is given by $d_Q^2(X,\tilde{X})$. Observing this distance relative with regard to the expected squared norm of X, we obtain the ratio

$$(1.21) \qquad \frac{E(d_Q^2(X,\tilde{X}))}{E\|X\|_Q^2}$$

which assumes values between 0 and 1. In order to be in line with the existing definitions of goodness-of-fit measures, the ratio in (1.21) is subtracted from 1, yielding the coefficient

$$(1.22) \qquad \rho_Q^2 = 1 - \frac{E(d_Q^2(X,\tilde{X}))}{E\|X\|_Q^2}.$$

Let us first hypothesize that the goodness-of-fit measure assumes the extreme values 0 or 1. $\rho_Q^2 = 0$ implies that $E(d_Q^2(X,\tilde{X})) = E\|X\|_Q^2$, hence X is orthogonal to the best fitting hyperplane. This means that the vector $\tilde{X}$, for which the linear relation $B'\tilde{X} = 0$ holds exactly, is in no way related with X; there is not even a slight indication of a linear association between the components of X. When ρ_Q^2 is equal to 1, then X and $\tilde{X}$ coincide and the hypothesized linear relation holds exactly for X. Hence, intuitively one may think of ρ_Q^2 as a quantity that measures the fit of the linear model. Furthermore, utilizing this definition of goodness-of-fit, it is not difficult to obtain expressions for ρ_Q^2 when a specific type of the PR model (like Simultaneous Equations (SE) or Seemingly Unrelated Regressions (SUR)) is analysed, as the expected squared distance $E(d_Q^2(X,\tilde{X}))$ has already been analysed thoroughly in order to derive expressions for the optimal values of B (like B_{se} and B_{sur}).

So far we have utilized the definition of "length" to describe the minimum distance parameters. Also the goodness-of-fit measure definition is based on this concept of metric spaces. In fact, the minimum distance parameters and ρ_Q^2 are closely related, since the optimal value of B that minimizes the mean squared distance between $\tilde{X}$ and X, also maximizes the corresponding goodness-of-fit measure (see (1.22)). Before we go on by specifying ρ_Q^2 for various values of B, a second basic concept will be defined, i.e. the angle between two vectors. When x and ξ are two vectors in an Euclidean space with the metric defined by Q, then the cosine of the angle between x and ξ (denoted by $\underline{/}(x,\xi)$) is given by the ratio

$$(1.23) \qquad \cos \underline{/}(x,\xi) = \frac{x'Q\xi}{\|x\|_Q\|\xi\|_Q} = \frac{x'Q\xi}{\sqrt{(x'Qx)}\sqrt{(\xi'Q\xi)}}.$$

Some important properties of this measure are

(1.24) $-1 \leq \cos \angle(x,\xi) \leq 1$

$\cos \angle(x,\xi) = 0 \iff x \perp_Q \xi$

$(\cos \angle(x,\xi) = -1$ or $\cos \angle(x,\xi) = 1) \iff x = c_o\xi$ for some scalar
constant c_o.

More theory on angles between vectors in metric spaces can be found in Shilov (1977, chapter 8) and Pollock (1979, chapter 3). This metric concept can be used to elucidate some intuitive ideas behind the definition of ρ_Q^2. However, before the cosine definition is applied to the vectors X and $\tilde{X}$, one should realize that the angle formed by X and $\tilde{X}$ is stochastic in character. Therefore $E(X'Q\tilde{X})$ will be used, rather than $X'Q\tilde{X}$, as the definition of the inner product of X and $\tilde{X}$ (see also Luenberger (1969)). Analogous to (1.23), the cosine of the angle between the random vectors is given by

(1.25) $\cos \angle(X,\tilde{X}) = \dfrac{E(X'Q\tilde{X})}{\sqrt{E(X'QX)}\ \sqrt{E(\tilde{X}'Q\tilde{X})}}$,

which yields

(1.26) $\cos^2 \angle(X,\tilde{X}) = \dfrac{E^2(X'Q\tilde{X})}{E(X'QX)\ E(\tilde{X}'Q\tilde{X})}$.

Let us now compare the definition of $\cos^2$ in (1.26) with the definition of ρ_Q^2 in (1.22). From section III.1 it is known that $\tilde{X}$ is a linear transformation of X given by $(I_k - Q^{-1}B(B'Q^{-1}B)^{-1}B'X$ (see (III.1.15)). Substituting this expression for $\tilde{X}$ we obtain

(1.27) $X'Q\tilde{X} = \tilde{X}'QX = \tilde{X}'Q\tilde{X} = X'QX - X'B(B'Q^{-1}B)^{-1}B'X.$

Using the equalities of (1.27), the definition of ρ_Q^2 can be rewritten as

(1.28) $\quad \rho_Q^2 = 1 - \dfrac{E(X-\tilde{X})'Q(X-\tilde{X})}{E(X'QX)} = \dfrac{E(X'Q\tilde{X})}{E(X'QX)}.$

From multiplication of the numerator of (1.28) by $E(X'Q\tilde{X})$, multiplication of the denominator by $E(\tilde{X}'Q\tilde{X})$ and (1.27) it follows that

(1.29) $\quad \rho_Q^2 = \cos^2 \underline{/}(X,\tilde{X}).$

Hence, the goodness-of-fit coefficient can be interpreted as the squared cosine of the angle between X and its projection. Moreover, from (1.27) it follows that (1.28) can be rewritten as

(1.30) $\quad \rho_Q^2 = \dfrac{E(\tilde{X}'Q\tilde{X})}{E(X'QX)} = \left(\dfrac{E\|\tilde{X}\|_Q}{E\|X\|_Q}\right)^2$

from which it follows that the definition of ρ_Q^2 agrees with the usual way of defining the (squared) cosine of an angle in a rectangular triangle. Similar to the preceding derivation of ρ_Q^2, an expression for the goodness-of-fit measure in the PF notation can be obtained. That is, $\tilde{X}$ is replaced by $\hat{X}$ which equals according to (IV.1.12) $A(A'QA)^{-1}A'QX$, and the relevant formulas for ρ_Q^2 are given by

(1.31) $\quad \rho_Q^2 = 1 - \dfrac{E(d_Q^2(X,\hat{X}))}{E\|X\|_Q^2} = \left(\dfrac{E\|\hat{X}\|_Q}{E\|X\|_Q}\right)^2.$

the remaining sections of this chapter ρ_Q^2 is examined for the various forms B and A, that have been analysed in the chapters III and IV. It will appear t the coefficients of linear association are, similar to the optimal values the parameter matrices B and A, differentiable functions in Σ. Hence, timators of the various goodness-of-fit coefficients and their large sample properties can be obtained by means of the method of moments and the delta method.

V.2 Coefficients of linear association for principal relations and principal factors

In section 1 a general set-up has been given for a goodness-of-fit coefficient that measures the association between X and $\tilde{X}$ (or $\hat{X}$) on a scale from 0 to 1, yielding ρ_Q^2 (see (1.22)) which is geometrically interpreted as a squared cosine. Expressions for $\tilde{X}$ and $\hat{X}$ are in PR and in PF analysis obtained as solutions of a composite distance minimization problem. The first minimization yields the vectors $\tilde{X}$ and $\hat{X}$, that have minimum Q-distance to X and fulfill the p linear relations $B'\tilde{X} = 0$ or q linear relations $\hat{X} = AU$ respectively. Replacing $\tilde{X}$ and $\hat{X}$ in ρ_Q^2 by their minimum distance solutions, taking expectations and using trace operator properties, the goodness-of-fit measure can be rewritten as

(2.1) $\qquad \rho_{pr}^2 = 1 - \dfrac{tr((B'Q^{-1}B)^{-1}B'\Sigma B)}{tr(Q\Sigma)}$,

or

(2.2) $\qquad \rho_{pf}^2 = 1 - \dfrac{tr(Q\Sigma - (A'QA)^{-1}A'Q\Sigma QA)}{tr(Q\Sigma)} - \dfrac{tr((A'QA)^{-1}A'Q\Sigma QA)}{tr(Q\Sigma)}$.

The goodness-of-fit measures ρ_{pr}^2 and ρ_{pf}^2 will be referred to as coefficients of linear association for PR and PF respectively, as they indicate to what extent the linear relations $B'\tilde{X} = 0$ and $\hat{X} = AU$, that are derived for $\tilde{X}$ and $\hat{X}$, also hold for X itself. The second minimization yields values for B and A such that the expectation of the squared distances, used in the numerators of (2.1) and (2.2), are minimized. It is known from sections III.1 and IV.1 that the optimal values of B and A are defined by means of the characteristic vectors of covariance matrix Σ. In the following some important PR and PF results are brought to mind.

Let $\lambda_1 > \lambda_2 > \ldots > \lambda_k$ be the characteristic roots of Σ in the Euclidean metric defined by Q^{-1}. We define the following diagonal matrices

(2.3) $\qquad \Lambda_k = diag(\lambda_1,\ldots,\lambda_k)$

$\qquad\qquad \Lambda_{min,p} = diag(\lambda_{k-p+1},\ldots,\lambda_k)$

$\qquad\qquad \Lambda_{max,q} = diag(\lambda_1,\ldots,\lambda_q)$.

The optimal values of B and A, indicated by B_{pr} and A_{pf} respectively, are given by

(2.4) $\quad \Sigma B_{pr} = Q^{-1} B_{pr} \Lambda_{min,p}$

and

(2.5) $\quad \Sigma Q A_{pf} = A_{pf} \Lambda_{max,q}$.

Substitution of (2.4) and (2.5) in (2.1) and (2.2) yields

(2.6) $\quad \rho_{pr}^2 = 1 - \dfrac{\text{tr } \Lambda_{min,p}}{\text{tr}(Q\Sigma)}$

and

(2.7) $\quad \rho_{pf}^2 = \dfrac{\text{tr } \Lambda_{max,q}}{\text{tr}(Q\Sigma)}$.

As the trace of matrix $Q\Sigma$ equals the sum of its characteristic roots, we have

(2.8) $\quad \text{tr}(Q\Sigma) = \text{tr } \Lambda_k$,

hence replacing $\text{tr}(Q\Sigma)$ by $\text{tr } \Lambda_k$, (2.6) and (2.7) can be rewritten as

(2.9) $\quad \rho_{pr}^2 = \sum_{i=1}^{k-p} \lambda_i \Big/ \sum_{i=1}^{k} \lambda_i$

and

(2.10) $\quad \rho_{pf}^2 = \sum_{i=1}^{q} \lambda_i \Big/ \sum_{i=1}^{k} \lambda_i$.

What is the interpretation of these coefficients of linear association? The denominator $\text{tr } \Lambda_k$ represents the total amount of variation in the components of X. Approximating X by $\tilde{X}$ or $\hat{X}$ implies a neglect of variance in X in p and k-q dimensions respectively. The smallest characteristic roots indicate the amount of variance in X, not caught by its approximation. Consequently,

ρ^2_{pr} and ρ^2_{pf} stand for the relative amount of variance in X preserved by the dimension reduction which is implied by the restriction $B'\tilde{X} = 0$ or $\hat{X} = AU$. Note that this exactly describes the major intention of principal components analysis.

The MME $\hat{\rho}^2_{pr}$ and $\hat{\rho}^2_{pf}$ of ρ^2_{pr} and ρ^2_{pf} are obtained by replacing in the aforementioned expressions the population covariance matrix Σ by its sample counterpart. It is assumed that Σ is consistently estimated by $\hat{\Sigma}$. In order to make large sample inference statements about the estimators, it is also assumed that $\sqrt{N}(\text{vec}\hat{\Sigma} - \text{vec}\Sigma)$ is asymptotically normally distributed with mean 0 and asymptotic covariance matrix V. Hence when $\hat{\lambda}_1 > \hat{\lambda}_2 > \ldots > \hat{\lambda}_k$ are defined as the characteristic roots of $\hat{\Sigma}$ in the Q^{-1}-metric, then

(2.11) $\qquad \hat{\rho}^2_{pr} = \sum\limits_{i=1}^{k-p} \hat{\lambda}_i \, / \, \sum\limits_{i=1}^{k} \hat{\lambda}_i$

and

(2.12) $\qquad \hat{\rho}^2_{pf} = \sum\limits_{i=1}^{q} \hat{\lambda}_i \, / \, \sum\limits_{i=1}^{k} \hat{\lambda}_i$

determine consistent and asymptotically normally distributed estimators of the coefficients of linear association. Their limiting distributions are given by

(2.13) $\qquad \sqrt{N}(\hat{\rho}^2_{pr} - \rho^2_{pr}) \xrightarrow{D} N(0, \, [\nabla(\rho^2_{pr};\Sigma)]V[\nabla(\rho^2_{pr};\Sigma)]')$

and

(2.14) $\qquad \sqrt{N}(\hat{\rho}^2_{pf} - \rho^2_{pf}) \xrightarrow{D} N(0, \, [\nabla(\rho^2_{pf};\Sigma)]V[\nabla(\rho^2_{pf};\Sigma)]')$.

The only difficulty left is the derivation of the partial derivatives denoted by $\nabla(\rho^2_{pr};\Sigma)$ and $\nabla(\rho^2_{pf};\Sigma)$. The computation of these row vectors will be provided in two stages. Firstly, the derivative of a characteristic root λ_i of Σ (in the metric defined by Q^{-1}) with respect to $(\text{vec}\Sigma)'$ will be determined. Corresponding results for characteristic roots in the Euclidean metric defined by I_k have been derived before, for instance by Laughton (1964), Gandolfo (1981) and Phillips (1982). They have formulated derivatives of simple characteristic roots. What happens when we have a characteristic root with

multiplicity greater than one? Let us, for simplicity, restrict ourselves to a
(2×2) matrix $\Sigma=[\sigma_{ij}]$, where $i,j = 1,2$. The two characteristic roots λ_1 and λ_2
of Σ are defined as the roots of the characteristic equation, which is given by
$|\Sigma-\lambda I_2| = \lambda^2 - (\sigma_{11}+\sigma_{22})\lambda + (\sigma_{11}\sigma_{22}-\sigma_{21}\sigma_{12}) = 0$. For the partial derivative
of λ_1 with respect to σ_{11}, this characteristic root is considered to be a
function of σ_{11}, with parameters σ_{21}, σ_{12} and σ_{22}. That is $\lambda_1 =$
$\lambda_1(\sigma_{11};\sigma_{21},\sigma_{12},\sigma_{22})$, for $i=1,2$. A sketch of λ_1 and λ_2 as functions of σ_{11} is
presented in figure V.2.1.

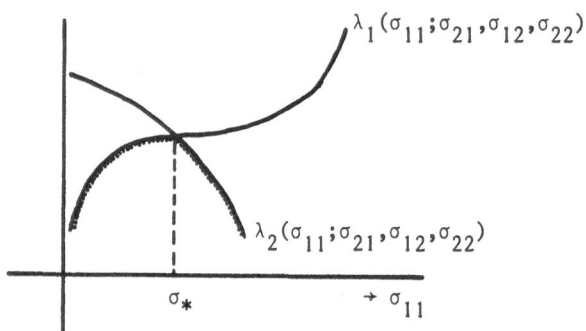

Figure V.2.1 Two characteristic roots λ_1 and λ_2 as functions of σ_{11}, with
parameters σ_{21}, σ_{12}, σ_{22}.

Let σ_* be the value of σ_{11} for which $\lambda_1 = \lambda_2$. Note that for $\sigma_{11} < \sigma_*$ the
smallest characteristic root is, in the figure, given by λ_1, while for $\sigma_{11} >$
σ_* the smallest characteristic root is given by λ_2. For $\sigma_{11} = \sigma_*$ we find that
Σ has one characteristic root with multiplicity two. Clearly the smallest
characteristic root (the dotted line in figure V.2.1) is not differentiable in
σ_*.
Let us return to the case of a square matrix Σ of order k. For the
differentiability of the characteristic root λ_i, its multiplicity must equal
one. Therefore it is assumed that Σ has k distinct characteristic roots
$\lambda_1 > \lambda_2 > \ldots > \lambda_k$.

In the second stage, the derivative of λ_1 will be used to obtain expressions
for the partial derivatives in question. Clearly, as the value
of $\rho_{pr_2}^2$ (ρ_{pf}^2) lies in the interval [0,1], some complications may be expected
when $\rho_{pr}^2=0$ $(\rho_{pf}^2=0)$ or $\rho_{pr}^2=1$ $(\rho_{pf}^2=1)$. However, these extreme values are
excluded from the set of possible values, as it is assumed that Σ is a

positive definite (hence non-singular) matrix. That is, its characteristic roots are all greater than 0.

Together with the assumption of distinct characteristic roots this implies $0 < \rho_{pr}^2 < 1$ (and $0 < \rho_{pf}^2 < 1$). The aforementioned two stages will be dealt with separately in the following two theorems.

<u>Theorem</u> (the derivative of the characteristic roots)
Let λ and β_{pr} be a characteristic root and the corresponding uniquely defined, normalized characteristic vector of Σ in the Q^{-1}-metric, that is

(2.15) $\qquad \Sigma\beta_{pr} = Q^{-1}\beta_{pr}\lambda \quad \text{and} \quad \beta_{pr}'Q^{-1}\beta_{pr} = 1.$

Then λ may be considered as a differentiable function of Σ with

(2.16) $\qquad \nabla(\lambda;\Sigma) = (\beta_{pr} \otimes \beta_{pr})' \qquad\qquad\qquad (1 \times k^2).$

The preceding theorem is given in terms of PR vector β_{pr}. The corresponding formulas for the PF case, where $Q\alpha_{pf}$ is a characteristic vector of Σ in the Q^{-1}-metric, are

(2.17) $\qquad \nabla(\lambda;\Sigma) = \frac{1}{\lambda}(Q\alpha_{pf} \otimes Q\alpha_{pf})' = \frac{1}{\lambda}(\alpha_{pf} \otimes \alpha_{pf})'(Q \otimes Q) \qquad (1 \times k^2).$

A factor $\frac{1}{\lambda}$ is introduced, as the characteristic vector $Q\alpha_{pf}$ is normalized in such a way that $(Q\alpha)'Q^{-1}(Q\alpha) = \lambda.$

<u>Proof</u>
From (2.15) it follows that

(2.18) $\qquad \lambda = \beta_{pr}'\Sigma\beta_{pr}$

and application of the product rule for matrix derivatives yields

(2.19) $\nabla(\lambda;\Sigma) = \dfrac{\partial\beta'_{pr}}{\partial(vec\Sigma)'} \; (\Sigma\beta_{pr} \; \boxtimes \; I_{k^2})$

$+ \; \beta'_{pr} \; \dfrac{\partial\Sigma}{\partial(vec\Sigma)'} \; (\beta_{pr} \; \boxtimes \; I_{k^2}) + \beta'_{pr}\Sigma \; \dfrac{\partial\beta_{pr}}{\partial(vec\Sigma)'}.$

Rewriting the first term at the right-hand side of (2.19) as

(2.20) $(vec(\dfrac{\partial\beta_{pr}}{\partial(vec\Sigma)'})')' \; (\Sigma\beta_{pr} \; \boxtimes \; I_{k^2})$

and using (A.11) it follows that (2.20) is equal to the third term at the right-hand side of (2.19), which can be rewritten as $\beta'_{pr}\Sigma \; \nabla(\beta_{pr};\Sigma)$, yielding

(2.21) $\nabla(\lambda;\Sigma) = \beta'_{pr} \; \dfrac{\partial\Sigma}{\partial(vec\Sigma)'} \; (\beta_{pr} \; \boxtimes \; I_{k^2}) + 2\beta'_{pr}\Sigma \; \nabla(\beta_{pr};\Sigma).$

From section III.1 it is known that $\nabla(\beta_{pr};\Sigma)$ equals $-(\beta'_{pr} \; \boxtimes \; Q^{\frac{1}{2}}(\Sigma^*-\lambda I_k)^+Q^{\frac{1}{2}})$, where $Q^{\frac{1}{2}}$ was defined as a symmetric ($k\times k$) matrix, such that

(2.22) $Q^{\frac{1}{2}}Q^{\frac{1}{2}} = Q$ and $Q^{\frac{1}{2}}Q^{-1}Q^{\frac{1}{2}} = I_k$

(see (III.1.11)). Hence, (2.15) can be rewritten as

(2.23) $\Sigma^*\beta^*_{pr} = \beta^*_{pr}\lambda$ and $\beta^{*\prime}_{pr}\beta^*_{pr} = 1$

with the transformed parameters defined by

(2.24) $\Sigma^* = Q^{\frac{1}{2}}\Sigma Q^{\frac{1}{2}}$ and $\beta^*_{pr} = Q^{-\frac{1}{2}}\beta_{pr}.$

Let us now return to (2.21). Replacing $\beta'_{pr}\Sigma$ by $\lambda\beta'_{pr}Q^{-1}$, the second term at the right-hand side of (2.21) is rewritten as

(2.25) $\qquad -2\lambda\beta'_{pr}Q^{-1}(\beta'_{pr} \otimes Q^{\frac{1}{2}}(\Sigma^*-\lambda I_k)^+Q^{\frac{1}{2}}) = -2\lambda(\beta'_{pr} \otimes \beta^{*\prime}_{pr}(\Sigma^*-\lambda I_k)^+Q^{\frac{1}{2}})$

which equals zero, as, according to (A.25), from $(\Sigma^*-\lambda I_k)\beta^*_{pr} = 0$ it follows that also $(\Sigma^*-\lambda I_k)^+\beta^*_{pr} = 0$.

The first term at the right-hand side of (2.21) remains and according to (A.47) it can be rewritten as

(2.26) $\qquad \nabla(\lambda;\Sigma) = \beta'_{pr}((\mathrm{vec}I_k)' \otimes I_k)(\beta_{pr} \otimes I_{k^2}).$

After application of some elementary Kronecker produkt properties, (2.25) results in

(2.27) $\qquad \nabla(\lambda;\Sigma) = (\mathrm{vec}(\beta_{pr}\beta'_{pr})) = (\beta_{pr} \otimes \beta_{pr})'. \quad \diamond$

As the coefficient of linear association ρ^2_{pr} is shown to be a simple function of the characteristic roots of Σ (see (2.9)), the matrix $\nabla(\rho^2_{pr};\Sigma)$ of partial derivatives follows from an application of the preceding theorem. Results are given in the following theorem.

Theorem (the derivatives of the PR coefficient of linear association)

Let ρ^2_{pr} be defined as denoted in (2.9), then its gradient matrix is given by

(2.28) $\qquad \nabla(\rho^2_{pr};\Sigma) = \dfrac{(\mathrm{vec}((1-\rho^2_{pr})Q - B_{pr}B'_{pr}))'}{\mathrm{tr}\ \Lambda_k} \qquad (1 \times k^2)$

with the diagonal elements of Λ_k and the columns of B_{pr} denoting the k characteristic roots and the p characteristic vectors (corresponding with the p minimal characteristic roots) of Σ in the Euclidean metric defined by Q^{-1}.

Proof

Considering ρ^2_{pr} as 1 minus a ratio of two functions in Σ, we may write

$$(2.29) \quad \nabla(\rho_{pr}^2;\Sigma) = \frac{\partial(1 - \frac{\text{tr } \Lambda_{min,p}}{\text{tr}(Q\Sigma)})}{\partial(\text{vec}\Sigma)'}$$

$$= - \frac{\text{tr}(Q\Sigma) \frac{\partial(\text{tr } \Lambda_{min,p})}{\partial(\text{vec}\Sigma)'} - (\text{tr } \Lambda_{min,p}) \frac{\partial \text{ tr}(Q\Sigma)}{\partial(\text{vec}\Sigma)'}}{(\text{tr}(Q\Sigma))^2}$$

$$= \frac{(1-\rho_{pr}^2)(\text{vec}(\frac{\partial \text{tr}(Q\Sigma)}{\partial \Sigma}))' - \sum_{i=1}^{p} (\beta_{pr,i} \otimes \beta_{pr,i})'}{\text{tr}(Q\Sigma)}$$

where the last equality follows from (2.16) with $\beta_{pr,i}$ the ith column of B_{pr}. Using matrix differentiation rule (A.49) for the trace operator and some elementary matrix properties, (2.29) can be rewritten as

$$(2.30) \quad \nabla(\rho_{pr}^2;\Sigma) = \frac{(1-\rho_{pr}^2)(\text{vec}Q)' - \sum_{i=1}^{p} (\text{vec}(\beta_{pr,i}\beta'_{pr,i}))'}{\text{tr}(Q\Sigma)}$$

$$= \frac{(1-\rho_{pr}^2)(\text{vec}Q)' - (\text{vec}(B_{pr}B'_{pr}))'}{\text{tr}(Q\Sigma)}.$$

Substitution of tr Λ_k for tr$(Q\Sigma)$ yields (2.28). ◇

A corresponding result can be given for the coefficient of linear association in the PF case. However, now the coefficient ρ_{pf}^2 is observed as a function of the maximal characteristic roots of Σ, as it is preferred to obtain the formulas in terms of matrix A_{pf} (the q characteristic vectors of Σ corresponding with the maximal characteristic roots, pre-multiplied by Q) rather than in terms of minimal characteristic vectors denoted by B_{pr}.

__Theorem__ (the derivative of the PF coefficient of linear association)
Let ρ_{pf}^2 be defined as denoted in (2.10), then its gradient matrix is given by

(2.31) $\quad V(\rho_{pf}^2;\Sigma) = \dfrac{(vec(QA_{pf}\Lambda_{max,q}^{-1}A_{pf}'Q - \rho_{pf}^2 Q))'}{tr\ \Lambda_k}$ $\qquad$ $(1\times k^2)$

with the diagonal elements of $\Lambda_{max,q}$ and the columns of QA_{pf} denoting the q maximal characteristic roots and the corresponding characteristic vectors of Σ in the Euclidean metric defined by Q^{-1}.

The proof of this theorem is similar to the derivation of $V(\rho_{pr}^2;\Sigma)$ and is left to the reader.

Let us return to the limiting distribution of $\hat\rho_{pr}^2$ and $\hat\rho_{pf}^2$, denoted in (2.13) and (2.14). The middle matrix V in the asymptotic variance represents the sample design. In the ideal-sample case of N i.i.d. observations, the matrix V equals the covariance matrix of $vec(XX')$ given by

(2.32) $\quad E(vec(XX'))(vec(XX'))' - (vec\Sigma)(vec\Sigma)'$ $\qquad$ $(k^2\times k^2)$.

A simple expression for the ideal-sample asymptotic variance of the PR coefficient of linear association is obtained by performing the pre- and post-multiplication, with V replaced by the matrix denoted in (2.32).

Theorem ($\hat\rho_{pr}^2$ for random i.i.d. sample observations)
Let $\{X_n: n=1,\ldots,N\}$ be a sample of N i.i.d. random observations on X and let $\hat\rho_{pr}^2$ be the MME of ρ_{pr}^2, that is defined as the coefficient of linear association in PR analysis (see (2.9)). Then

(2.33) $\quad \sqrt{N}\ (\hat\rho_{pr}^2 - \rho_{pr}^2) \overset{D}{\to} N(0,\ \dfrac{E[(1-\rho_{pr}^2)X'QX - X'B_{pr}B_{pr}'X]^2}{(tr\ \Lambda_k)^2}\ )$

with Λ_k and B_{pr} given in (2.3) and (2.4) and B_{pr} normalized as

(2.34) $\quad B_{pr}'Q^{-1}B_{pr} = I_p$.

Proof
Substituting for $V(\rho_{pr}^2;\Sigma)$ and V the matrix expressions denoted in (2.28) and (2.32) respectively, we get

(2.35) $\quad [\nabla(\rho_{pr}^2;\Sigma)]V[\nabla(\rho_{pr}^2;\Sigma)]'$

$$= \frac{(vec((1-\rho_{pr}^2)Q-B_{pr}B'_{pr}))'\ E(vec(XX'))(vec(XX'))'(vec((1-\rho_{pr}^2)Q-B_{pr}B'_{pr}))}{(tr\ \Lambda_k)^2}$$

$$- \frac{(vec((1-\rho_{pr}^2)Q-B_{pr}B'_{pr}))'\ (vec\Sigma)(vec\Sigma)'\ (vec((1-\rho_{pr}^2)Q-B_{pr}B'_{pr}))}{(tr\ \Lambda_k)^2}.$$

Note that both the first and second term at the right hand side of (2.35) are scalars. Using (A.14) this kind of scalars can be rewritten as the trace of a matrix product, yielding

(2.36) $\quad \dfrac{E[tr(((1-\rho_{pr}^2)Q-B_{pr}B'_{pr})XX')]^2}{(tr\ \Lambda_k)^2} - \dfrac{[(1-\rho_{pr}^2)tr(Q\Sigma)\ -\ tr(B'_{pr}\Sigma B_{pr})]^2}{(tr\ \Lambda_k)^2}.$

Replacing $tr(Q\Sigma)$ by $tr\ \Lambda_k$ (see (2.8)) and $tr(B'_{pr}\Sigma B_{pr})$ by $tr\ \Lambda_{min,p}$ (which is allowed because of (2.34)), the second term at the right hand side of (2.36) can be shown to equal zero. Using elementary trace properties the remaining non-zero part of the expression is shown to equal the variance of the limiting distribution denoted in (2.33). ◊

In a similar way the ideal-sample limiting distribution of $\hat\rho_{pf}^2$ can be derived, by substituting for $\nabla(\rho_{pf}^2;\Sigma)$ in the variance of limiting distribution (2.14), the matrix expression denoted in (2.31). The results are stated, without proof, in the following theorem.

Theorem ($\hat\rho_{pf}^2$ for random i.i.d. sample)
Let $\{X_n: n=1,\ldots,N\}$ be a sample of N i.i.d. random observations on X and let $\hat\rho_{pf}^2$ be the MME of ρ_{pf}^2, that is defined as the coefficient of linear association in the PF analysis (see (2.10)). Then

$$(2.37) \qquad \sqrt{N} \ (\hat{\rho}_{pf}^2 - \rho_{pf}^2) \xrightarrow{D} N(0, \ \frac{E[X'QA_{pf}\Lambda_{max,q}^{-1}A'_{pf}QX - \rho_{pr}^2 X'QX]^2}{(tr \ \Lambda_k)^2} \ )$$

with Λ_k and A_{pf} given in (2.3) and (2.5) and A_{pf} normalized as

$$(2.38) \qquad A'_{pf}QA_{pf} = \Lambda_{max,q}.$$

In the preceding sections we have recognized for both the PF and the PR case a number of multivariate techniques as special cases. This section ends by discussing two of these types, viz. Principal Components (PC) analysis and Canonical Correlation (CC) analysis, yielding simplified expressions for the PF and PR versions of the coefficients of linear association.

In section IV.3 it has been shown that, setting the metric defining matrix Q equal to I_k and working with standardized observations, such that Σ is replaced by correlation matrix R, the results obtained by PF analysis equal the coefficients of the principal components. The purpose of PC analysis is to characterize or explain the variability in vector X, by replacing it by a linear combination of a smaller number of mutually independent (principal) components with large variance. The extent to which the total variance is explained by means of a less dimensional vector is indicated by the coefficient of linear association ρ_{pc}^2, which is obtained as a special case of ρ_{pr}^2, and is given by

$$(2.39) \qquad \rho_{pc}^2 = \frac{tr(\Lambda_{max,q})}{tr \ R} = \frac{1}{k} \sum_{i=1}^{q} \lambda_i$$

where λ_i now denotes the ith characteristic root of correlation matrix R in the simple Euclidean metric defined by I_k. As k denotes the total variance of the standardized components of X and λ_i the variance of the ith principal component, ρ_{pc}^2 stands for the fraction of the total variance represented by the set of q principal components.

In order to obtain large sample properties of the MME of ρ_{pc}^2, the partial derivatives of ρ_{pc}^2 are needed. Having knowledge of the derivatives of the characteristic roots of a covariance matrix, it is not that difficult to prove that the gradient of ρ_{pc}^2 is given by

$$(2.40) \qquad \nabla(\rho_{pc}^2;\Sigma) = \frac{1}{k} (vec(A_{pc}A'_{pc}))'$$

with A_{pc} the $(k \times q)$ matrix of PC parameters. Note the difference of this result with the gradient of ρ_{pc}^2, given in (2.31). $\nabla(\rho_{pc}^2;\Sigma)$ is not merely obtained by

replacing in $\nabla(\rho_{pc}^2;\Sigma)$ the matrix Q by I_k and tr Λ_k by k. Dissimilarities are caused by the fact that in the PC case the total variance in X is no longer a function of Σ, but equals k as a result of the standardization.

In the multivariate statistical analysis, asymptotic distributions of characteristic root estimators are used to test whether they are equal, or whether one or more of the characteristic roots are zero. When for instance the first hypothesis is accepted, then this implies that all principal components have the same variance and, hence, it is useless to reduce the dimension of X by applying PC analysis. In the latter case the zero variances indicate succesful use of the PC technique. A comprehensive review of the literature on inference problems in the PC analysis has been given by Muirhead (1982, chapter 9).

The second case that will be distinghuished is a special case of PR analysis. In section III.6 it has been shown that partitioning X in a Y and a Z component

and setting the metric defining matrix Q equal to $Q_{cc} = \begin{bmatrix} \Sigma_{yy}^{-1} & 0 \\ \hline 0 & \Sigma_{zz}^{-1} \end{bmatrix}$, the

results obtained by PR analysis are similar to the coefficients of the canonical variables. The purpose of CC analysis is to derive linear combinations of Y and Z with a maximum correlation. The choice of the metric matrix Q_{cc} yields the coefficient ρ_{pr}^2 to be modified into a coefficient of linear association, say ρ_{cc}^2, that indicates the total correlation between the canonical variables. From (2.6) and the definition of the canonical correlations it follows that

$$(2.41) \qquad \rho_{cc}^2 = 1 - \frac{1}{k} \text{tr}(\Lambda_{cc,min,p}) = 1 - \frac{1}{k} \sum_{i=1}^{p} (1-\phi_i)$$

where $\Lambda_{cc,min,p}$ is the diagonal matrix with the maximum characteristic roots of Σ in the Q_{cc}^{-1} metric as its diagonal elements. The scalar parameters $\phi_1,\ldots,\phi_p$ denote the p canonical correlations between Y and Z. Replacing the population moments by sample moments, the MME of ρ_{cc}^2 is obtained and the limiting distribution of this estimator can be determined when the partial derivatives of $\rho^2_{cc_1}$ with respect to $(\text{vec}\Sigma)'$ are known. Although the metric defining matrix Q^{-1}_{cc} itself depends on Σ, the derivative of the characteristic root λ_{cc} with respect to $(\text{vec}\Sigma)'$ remains the same as in the PR case, viz.

(2.42) $\qquad \nabla(\lambda_{cc};\Sigma) = (\beta_{cc} \otimes \beta_{cc})'.$

Hence, from the definition of ρ^2_{cc} (see (2.41)) it follows that

(2.43) $\qquad \nabla(\rho^2_{cc};\Sigma) = -\frac{1}{k} \sum_{i=1}^{p} (\beta_{cc,i} \otimes \beta_{cc,i})' = -\frac{1}{k} (\text{vec}(B_{cc}B'_{cc}))',$

with B_{cc} the $(k \times p)$ matrix of CC parameters.

The connection between the coefficient of linear association and the coefficient of multiple correlation has yet not been made. This will be postponed until the end of the next section, in which the vector X will be split in Y- and Z-components, like in section 1, to derive the coefficient of linear association for the structural simultaneous equations model.

V.3 Coefficients of linear association for simultaneous equations systems

The structural SE systems are in section III.3 introduced as a special case of PR analysis. For this technique the random vector X is partitioned into two components Y and Z, such that $X = (Y',Z')'$ with Y an (m×1) vector. The Z component represents the exogenous variables, that were defined as the variables of X that have to remain unchanged when X is projected onto the (k-p)-dimensional best fitting hyperplane. This means that the expected squared distance between X and $\tilde{X}$ is determined by the Y-dimension only and is given by $E(d_{Q_{yy}}^2 (Y,\tilde{Y}))$. Therefore it is logical to rewrite the goodness-of-fit measure, denoted in (1.22), in terms of the Y-dimensions only, yielding

$$(3.1) \qquad \rho_{Q_{yy}}^2 = 1 - \frac{E(Y-\tilde{Y})'Q_{yy}(Y-\tilde{Y})}{E(Y'Q_{yy}Y)}.$$

From section III.2 it is known that $\tilde{Y}$ is a projection of X given by $Y - Q_{yy}^{-1}B_y(B_y'Q_{yy}^{-1}B_y)^{-1}B'X$. Substitution of this expression in (3.1) yields a coefficient of linear association for the SE model. Taking expectations and using the trace operator, the three different factors of this coefficient can be written as

$$(3.2) \qquad E(Y'Q_{yy}Y) = tr(Q_{yy}\Sigma_{yy}),$$

$$(3.3) \qquad E(Y'Q_{yy}\tilde{Y}) = tr(Q_{yy}\Sigma_{yy}) - tr((B_y'Q_{yy}^{-1}B_y)^{-1}(B_y'\Sigma_{yy}B_y + B_z'\Sigma_{zy}B_y))$$

and

$$(3.4) \qquad E(\tilde{Y}'Q_{yy}\tilde{Y}) = tr(Q_{yy}\Sigma_{yy}) - tr((B_y'Q_{yy}^{-1}B_y)^{-1}(B_y'\Sigma_{yy}B_y - B_z'\Sigma_{zz}B_z)).$$

Recall that in section 1 it was shown that $E(X'Q\tilde{X}) = E(\tilde{X}'Q\tilde{X})$, which was used to simplify ρ_Q^2 as denoted in (1.30). From (3.3) and (3.4) it follows that in general $E(Y'Q_{yy}\tilde{Y}) \neq E(\tilde{Y}'Q_{yy}\tilde{Y})$, hence a corresponding simplified definition of the coefficient of linear association is not yet possible in the SE case. However, one should realize that the coefficient is evaluated for the B_{se} matrix that minimizes the mean squared distance $E(d_{Q_{yy}}^2 (Y,\tilde{Y}))$. It has been shown that partitioning B_{se} as $[B_{se,y}' \vdots B_{se,z}']'$, the two submatrices are linearly related as (see (III.3.13))

$$(3.5) \qquad B_{se,z} = -\Sigma_{zz}^{-1}\Sigma_{zy}B_{se,y}.$$

Replacing in (3.3) and (3.4) matrix B by its optimal value B_{se} and using (3.5) it can be seen that then $E(Y'Q_{yy}\tilde{Y}) = E(\tilde{Y}'Q_{yy}\tilde{Y})$. Hence, evaluation of the coefficient of linear association for the SE model, say ρ_{se}^2, in its optimal value yields

$$(3.6) \qquad B = B_{se} \Rightarrow \rho_{se}^2 = \frac{E(\tilde{Y}'Q_{yy}\tilde{Y})}{E(Y'Q_{yy}Y)} = \left(\frac{E\|\tilde{Y}\|_{Q_{yy}}}{E\|Y\|_{Q_{yy}}}\right)^2.$$

From (3.6) it follows that the definition of ρ_{se}^2 agrees with the definition of the squared cosine between two vectors in a rectangular triangle in an m-dimensional space spanned by the endogenous variables of X.

Substituting the optimal value $Y - Q_{yy}^{-1}B_y(B_y'Q_{yy}^{-1}B_y)^{-1}B'X$ for $\tilde{Y}$ in (3.1) we obtain

$$(3.7) \qquad \rho_{se}^2 = 1 - \frac{\mathrm{tr}((B_{se,y}'Q_{yy}^{-1}B_{se,y})^{-1}B_{se}'\Sigma B_{se})}{\mathrm{tr}(Q_{yy}\Sigma_{yy})}$$

and application of relation (3.5) yields

$$(3.8) \qquad \rho_{se}^2 = 1 - \frac{\mathrm{tr}((B_{se,y}'Q_{yy}^{-1}B_{se,y})^{-1}B_{se,y}'(\Sigma_{yy}-\Sigma_{zy}'\Sigma_{zz}^{-1}\Sigma_{zy})B_{se,y})}{\mathrm{tr}(Q_{yy}\Sigma_{yy})}.$$

From section III.3 it is also known (see (III.3.13)) that $B_{se,y}$ is given by

$$(3.9) \qquad (\Sigma_{yy}-\Sigma_{zy}'\Sigma_{zz}^{-1}\Sigma_{zy})B_{se,y} = Q_{yy}^{-1}B_{se,y}\Omega_{\min,p}$$

with $\Omega_{\min,p}$ the diagonal matrix with on its diagonal the p minimal characteristic roots of $(\Sigma_{yy}-\Sigma_{zy}'\Sigma_{zz}^{-1}\Sigma_{zy})$ in the Euclidean metric defined by Q_{yy}^{-1}. Substitution of (3.9) in (3.8) yields

$$(3.10) \qquad \rho_{se}^2 = 1 - \frac{\text{tr } \Omega_{\min,p}}{\text{tr}(Q_{yy}\Sigma_{yy})}.$$

Note that, again, the characteristic roots (in this case of matrix $(\Sigma_{yy}-\Sigma'_{zy}\Sigma_{zz}^{-1}\Sigma_{zy})$) play an important role in the determination of the goodness-of-fit of the structural linear model. However, the denominator equals $\text{tr}(Q_{yy}\Sigma_{yy})$, which is equal to the sum of the characteristic roots of matrix Σ_{yy} (and not of matrix $(\Sigma_{yy}-\Sigma'_{zy}\Sigma_{zz}^{-1}\Sigma_{zy})$) in the Q_{yy}^{-1}-metric.
Let us observe the extreme values of ρ_{se}^2. If it is assumed that the hypothesized linear relation tends to hold exactly, that is not only $\tilde{Z} = Z$, but also $\tilde{Y} = Y$, then

$$(3.11) \qquad B_{se,y}Y + B_{se,z}Z = 0.$$

Post-multiplying (3.11) by Y', taking expectations and replacing $B_{se,z}$ by $-\Sigma_{zz}^{-1}\Sigma_{zy}B_{se,y}$ we obtain

$$(3.12) \qquad B'_{se,y}(\Sigma_{yy}-\Sigma'_{zy}\Sigma_{zz}^{-1}\Sigma_{zy}) = 0.$$

As $B_{se,y}$ consists of p independent columns, this means that the rank of $(\Sigma_{yy}-\Sigma'_{zy}\Sigma_{zz}^{-1}\Sigma_{zy})$ is at most (m-p) and its p minimal characteristic roots are zero. Consequently, $\Omega_{\min,p}$ is the (p×p) zero matrix and ρ_{se}^2 is 1.
In the opposite case of two vectors Y and $\tilde{Y}$, that are orthogonal (in the Q_{yy}-metric), it follows that

$$(3.13) \qquad Y'Q_{yy}\tilde{Y} = Y'Q_{yy}Y - Y'B_{se,y}(B'_{se,y}Q_{yy}^{-1}B_{se,y})^{-1}(B'_{se,y}Y + B'_{se,z}Z) = 0.$$

Taking expectations, using the trace operator and applying (3.5), (3.13) can be rewritten as

$$(3.14) \qquad \text{tr}(Q_{yy}\Sigma_{yy}) - \text{tr}((B'_{se,y}Q_{yy}^{-1}B_{se,y})^{-1}B'_{se,y}(\Sigma_{yy}-\Sigma'_{zy}\Sigma_{zz}^{-1}\Sigma_{zy})B_{se,y}) = 0.$$

Substitution of (3.9) in (3.14) yields

(3.15) $\text{tr}(Q_{yy}\Sigma_{yy}) - \text{tr}(\Omega_{\min,p}) = 0,$

which implies that in case of orthogonality of the vectors Y and $\tilde{Y}$, the correlation of linear association for the SE model equals zero. Hence ρ_{se}^2 satisfies the usual properties, associated with a goodness-of-fit measure, so that this definition of a coefficient of linear association is intuitively appropriate. However, it should be noted that 0 and 1 are only hypothetical values for ρ_{se}^2. Similar as in the PR and PF case it can be argued from the positive definiteness of Σ_{yy} and $(\Sigma_{yy}-\Sigma_{zy}'\Sigma_{zz}^{-1}\Sigma_{zy})$, that $0 < \rho_{se}^2 < 1$.

Now we come to the estimation of ρ_{se}^2. Therefore it is assumed that we have knowledge of a consistent and asymptotically normally distributed estimator $\hat{\Sigma}$ of Σ, which is, similar to Σ, partitioned in a Y- part and a Z-part. Denoting by $\hat{\Omega}_{\min,p}$ the diagonal matrix with on its diagonal the p smallest characteristic roots of $(\hat{\Sigma}_{yy}-\hat{\Sigma}_{zy}'\hat{\Sigma}_{zz}^{-1}\hat{\Sigma}_{zy})$ in the metric defined by Q_{yy}^{-1}, the MME $\hat{\rho}_{se}^2$ of the SE coefficient of linear association is, analogous to (3.10) defined by

(3.16) $\hat{\rho}_{se}^2 = 1 - \dfrac{\text{tr } \hat{\Omega}_{\min,p}}{\text{tr}(Q_{yy}\hat{\Sigma}_{yy})}.$

In order to obtain the limiting distribution of $\hat{\rho}_{se}^2$ by means of the delta method, the derivative of ρ_{se}^2 with respect to $(\text{vec}\Sigma)'$ is needed. As the coefficient of linear association is determined by the characteristic roots of $(\Sigma_{yy}-\Sigma_{zy}'\Sigma_{zz}^{-1}\Sigma_{zy})$, first the derivative of such a characteristic root with respect to $(\text{vec}\Sigma)'$ will be given in the following theorem. For differentiability of the characteristic root, it is assumed that its multiplicity equals 1.

<u>Theorem</u> (derivative of the characteristic root of $(\Sigma_{yy}-\Sigma_{zy}'\Sigma_{zz}^{-1}\Sigma_{zy})$).
Let ω and $\beta_{se,y}$ be a characteristic root and the corresponding, uniquely defined normalized characteristic vector of $(\Sigma_{yy}-\Sigma_{zy}'\Sigma_{zz}^{-1}\Sigma_{zy})$ in the Q_{yy}^{-1}-metric. That is

(3.17) $(\Sigma_{yy}-\Sigma_{zy}'\Sigma_{zz}^{-1}\Sigma_{zy})\beta_{se,y} = Q_{yy}^{-1}\beta_{se,y}\omega$ and $\beta_{se,y}'Q_{yy}^{-1}\beta_{se,y} = 1.$

Then ω may be considered as a differentiable function of Σ with

$$(3.18) \qquad \nabla(\omega;\Sigma) = (\beta_{se} \otimes \beta_{se})' \qquad\qquad (1 \times k^2)$$

where

$$(3.19) \qquad \beta_{se} = (\beta'_{se,y}, \beta'_{se,z})' \quad \text{and} \quad \beta_{se,z} = -\Sigma_{zz}^{-1} \Sigma_{zy} \beta_{se,y}.$$

Rewriting characteristic root definition (3.17) as

$$(3.20) \qquad \Sigma\beta_{se} = Q_o^+ \beta_{se} \omega \quad \text{and} \quad \beta'_{se} Q_o^+ \beta_{se} = 1$$

$$\text{with } Q_o^+ = \begin{bmatrix} Q_{yy}^{-1} & \vdots & 0 \\ --- & \vdots & -- \\ 0 & \vdots & 0 \end{bmatrix} \qquad\qquad (k \times k)$$

the problem is reformulated, so that it corresponds with the characteristic root specification in the PR case (see (2.15)). See also section III.3, especially (III.3.18), where Q_o was introduced as a matrix, defining a semi-metric in $\mathbb{R}^k$. Defining the transformation matrix $Q^{\frac{1}{2}}$ (like in section III.3) by

$$(3.21) \qquad Q^{\frac{1}{2}} = \begin{bmatrix} Q_{yy}^{\frac{1}{2}} & \vdots & 0 \\ --- & \vdots & ---- \\ 0 & \vdots & I_{k-m} \end{bmatrix}$$

with $Q_{yy}^{\frac{1}{2}}$ a symmetric (m×m) matrix, in such a way that

$$(3.22) \qquad Q_{yy}^{\frac{1}{2}} Q_{yy}^{\frac{1}{2}} = Q_{yy} \quad \text{and} \quad Q_{yy}^{\frac{1}{2}} Q_{yy}^{-1} Q_{yy}^{\frac{1}{2}} = I_m,$$

the random vector X is transformed into $X^* = Q^{\frac{1}{2}}X$, so that the metric in its Y dimensions is defined by I_m and the component in the Z-dimensions remains unchanged. Hence for the proof of this theorem one is referred to the derivation of the characteristic root derivatives, with an adjusted transformation of the observation vector.

Having knowledge of the derivatives of the diagonal elements of $\Omega_{min,p}$, it is not difficult to obtain the gradient vector of ρ_{se}^2. The results are given in the following theorem.

__Theorem__ (the derivative of the SE coefficient of linear association)
Let ρ_{se}^2 be defined as denoted in (3.10), then its gradient vector is given by

$$(3.23) \qquad \nabla(\rho_{se}^2;\Sigma) = \frac{(vec((1-\rho_{se}^2)Q_o - B_{se}B_{se}'))'}{tr(Q_{yy}\Sigma_{yy})} \qquad (1 \times k^2)$$

with B_{se} the SE parameter matrix defined in (3.5) and (3.9), and

$$Q_o = \left[\begin{array}{c|c} Q_{yy} & 0 \\ \hline 0 & 0 \end{array}\right].$$

Note the correspondence between the gradient matrices denoted in (2.28) and in (3.23). The lay-out of these expressions is in both cases the same. The only differences are, apart from the different subscripts "pr" and "se", that the factor tr Λ_k and matrix Q in (2.28) are in (3.23) replaced by $tr(Q_{yy}\Sigma_{yy})$ and Q_o respectively. Recalling that tr $\Lambda_k = tr(Q\Sigma)$ and observing that $tr(Q_{yy}\Sigma_{yy}) = tr(Q_o\Sigma)$ it follows that, in fact, the only difference is in the substitution of Q_o for Q. In other words, to prove this proposition, it is sufficient to refer to the derivation of $\nabla(\rho_{pr}^2;\Sigma)$ in section 2.

Combining the asymptotic normality of $(vec\hat{\Sigma})$ and the gradient vector of ρ_{se}^2, denoted in (3.23), the delta method yields the limiting distribution of $\hat{\rho}_{se}^2$. That is, when V denotes the asymptotic covariance matrix of $\sqrt{N}(vec\hat{\Sigma} - vec\Sigma)$, then

$$(3.24) \qquad \sqrt{N}(\hat{\rho}_{se}^2 - \rho_{se}^2) \xrightarrow{D} N(0, [\nabla(\rho_{se}^2;\Sigma)]V[\nabla(\rho_{se}^2;\Sigma)]').$$

The effect of the ideal-sample assumption on the limiting distribution denoted in (3.24) is as follows

__Theorem__ ($\hat{\rho}_{se}^2$ for random i.i.d. sample observations)
Let $\{X_n: n=1,...,N\}$ be a sample of N i.i.d. random observations on X and let

$\hat{\rho}^2_{se}$ be the MME of ρ^2_{se}, that is defined as the coefficient of linear association in the SE analysis (see (3.10)). Then

$$(3.25) \qquad \sqrt{N}(\hat{\rho}^2_{se} - \rho^2_{se}) \xrightarrow{D}$$

$$N(0, \frac{E[(1-\rho^2_{se})Y'Q_{yy}Y - (Y-\Sigma'_{zy}\Sigma^{-1}_{zz}Z)'B_{se,y}B'_{se,y}(Y-\Sigma'_{zy}\Sigma^{-1}_{zz}Z)]^2}{(tr(Q_{yy}\Sigma_{yy}))^2})$$

with $B_{se,y}$ defined in (3.9) and normalized as $B'_{se,y}Q^{-1}_{yy}B_{se,y} = I_p$.

Proof

The asymptotic normality has already been obtained in (3.24) by means of the delta method. This leaves us with the computation of the variance of the limiting distribution. In (3.23) we have seen that $V(\rho^2_{se};\Sigma)$ can simply be obtained from the specification of $V(\rho^2_{pr};\Sigma)$ by substituting Q_o for Q. Consequently, the resulting asymptotic variance is obtained by replacing in the asymptotic variance of $\hat{\rho}^2_{pr}$ in the ideal-sample case (see (2.33)) tr Λ_k by $tr(Q_{yy}\Sigma_{yy})$ and Q and Q_o. When, moreover, X and $B_{se,z}$ are replaced by $(Y',Z')'$ and $-\Sigma^{-1}_{zz}\Sigma_{zy}B_{se,y}$, then the variance as denoted at the right-hand side of (3.25) follows automatically. ◊

A special case of the SE coefficient of linear association is obtained when the number of relations (p) is assumed to equal the number of endogenous variables (m). In this case the matrix $\Omega_{min,p}$ includes all the m characteristic roots of $(\Sigma_{yy}-\Sigma'_{zy}\Sigma^{-1}_{zz}\Sigma_{zy})$ and consequently

$$(3.26) \qquad p = m \Rightarrow tr(\Omega_{min,p}) = tr(Q_{yy}(\Sigma_{yy}-\Sigma'_{zy}\Sigma^{-1}_{zz}\Sigma_{zy})).$$

Substitution of the right-hand side of (3.26) in the coefficient definition (3.10) of ρ^2_{se} yields

$$(3.27) \qquad p = m \Rightarrow \rho^2_{se} = \frac{tr(Q_{yy}\Sigma'_{zy}\Sigma^{-1}_{zz}\Sigma_{zy})}{tr(Q_{yy}\Sigma_{yy})}.$$

Let us now compare the coefficient of linear association ρ_{se}^2 as denoted in (3.27), with the trace correlation coefficient $\bar{\rho}_{yz}^2$ of section 1. Recall that $\bar{\rho}_{yz}^2$ denotes a goodness-of-fit measure for the reduced form linear model and is defined by (see (1.13))

$$(3.28) \qquad \bar{\rho}_{yz}^2 = \frac{1}{m} \, \text{tr}(\Sigma_{yy}^{-1}(\Sigma_{zy}'\Sigma_{zz}^{-1}\Sigma_{zy})).$$

Rewriting ρ_{se}^2 (for $p = m$) with Q_{yy} replaced by I_m we obtain

$$(3.29) \qquad \rho_{se}^2 = \frac{\text{tr}(\Sigma_{zy}'\Sigma_{zz}^{-1}\Sigma_{zy})}{\text{tr}(\Sigma_{yy})}.$$

In words this means that both $\bar{\rho}_{yz}^2$ and ρ_{se}^2 for $p=m$ are functions of the two matrices Σ_{yy} and $\Sigma_{zy}'\Sigma_{zz}^{-1}\Sigma_{zy}$. The difference is only in the order of inverting a matrix or scalar, applying the trace operator and performing a (matrix-) multiplication. A factor $\frac{1}{m}$ is used in connection with $\bar{\rho}_{yz}^2$ in order to obtain values in the [0,1] interval. This is not necessary for ρ_{se}^2, as in this case the [0,1]-scale is provided by double application of the trace operator. More about the interpretation of this kind of coefficients of linear association will be given in the next section of this chapter, where the coefficients are determined in the context of the seemingly unrelated regressions system.

V.4 Coefficients of linear association for seemingly unrelated regressions

A set of seemingly unrelated regressions is obtained by observing a set of
simultaneous equations for which the number of relations (p) equals the number
of endogenous variables (m) and where the (square) parameter sub-matrix B_y is
set equal to $-I_p$. As a result we have a set of p linear relations, and each
equation contains only one endogenous variable with a coefficient equal to -1.
The Z-component of X (that we consists of the exogenous variables of the
system) remains unchanged under the distance minimization, i.e. $\tilde{Z} = Z$.
Approximating $B'X = 0$ by $B'\tilde{X} = 0$ with the restrictions $B_y = -I_p$ and $\tilde{Z} = Z$, we
directly obtain

$$(4.1) \qquad \tilde{Y} = B'_z Z.$$

In section III.4 it was shown that the value of B_z, that minimizes the
expected squared distance between Y and $\tilde{Y}$, is given by

$$(4.2) \qquad B_{sur,z} = \Sigma_{zz}^{-1} \Sigma_{zy}.$$

As in the SUR analysis the divergence of X from $\tilde{X}$ is in the Y-dimensions only,
the goodness-of-fit measure $\rho_{Q_{yy}}^2$ is defined similar to the one in the SE case,
as denoted in (3.1). From (4.1) and (4.2) it follows that the optimal value
of $\tilde{Y}$ in the SUR case is given by $\Sigma_{zy}' \Sigma_{zz}^{-1} Z$. Substitution of this value in
$\rho_{Q_{yy}}^2$ yields a coefficient of linear association for SUR, given by

$$(4.3) \qquad \rho_{sur}^2 = \frac{tr(Q_{yy} \Sigma_{zy}' \Sigma_{zz}^{-1} \Sigma_{zy})}{tr(Q_{yy} \Sigma_{yy})}.$$

Note that this expression has also been obtained for the coefficient of linear
association in the SE case for p=m (see (3.27)) and consequently the remarks
made for ρ_{se}^2 at the end of section 3 also apply here. When p=1, and hence the
vector Y degenerates into a random variable, we consider only one linear
approximation of Y (given X). This case can be interpreted as the structural
linear regression model, where one random variable is "explained" by a vector
of (k-1) random explanatory variables. In this case we can dispense with the
two trace operators in (4.5) as their arguments are no longer matrices.

The coefficient ρ_{sur}^2 is then defined as the ration of two scalars, hence

(4.4) $p = 1 \Rightarrow \rho_{sur}^2 = \dfrac{\Sigma'_{zy} \Sigma_{zz}^{-1} \Sigma_{zy}}{\Sigma_{yy}}.$

Note that ρ_{sur}^2 for $p=1$ equals the population analogue of the multiple correlation coefficient R^2, that is defined in the context of the functional linear regression analysis as a goodness-of-fit measure (see (1.6)). When $p=1$, the parameter matrix $B_{sur,z}$ is a $((k-1) \times 1)$ column vector, which will in the following be denoted by $\beta_{sur,z}$. Using $\beta_{sur,z} = \Sigma_{zz}^{-1} \Sigma_{zy}$, (4.4) can be rewritten as

(4.5) $p = 1 \Rightarrow \rho_{sur}^2 = \dfrac{\beta'_{sur,z} \Sigma_{zz} \beta_{sur,z}}{\Sigma_{yy}} = \dfrac{var(\beta'_{sur,z} Z)}{var(Y)}$

where "var" stands for taking the variance. Hence ρ_{sur}^2 can be interpreted as the ratio of the variance of $\tilde{Y} = \beta'_{sur,z} Z$ (the linear approximation of Y), to the variance of Y. In general

(4.6) $\rho_{sur}^2 = \dfrac{tr(var(B'_{sur,z} Z))}{tr(var(Y))}$

where "var" now denotes a $(p \times p)$ covariance matrix. Hence ρ_{sur}^2 is the ratio of the sum of the variances of the p Y-components. This is a natural extension of the population analogue of R^2 denoted in (4.5).

A consistent estimator of ρ_{sur}^2 is obtained, like in the other versions of the goodness-of-fit measures for the structural relations, by means of the method of moments. Defining $\hat{\Sigma}$ to be a consistent and asymptotically normally distributed estimator of Σ, with a similar partitioning, the MME $\hat{\rho}_{sur}^2$ of ρ_{sur}^2 is given by

(4.7) $\hat{\rho}_{sur}^2 = \dfrac{tr(Q_{yy} \hat{\Sigma}'_{zy} \hat{\Sigma}_{zz}^{-1} \hat{\Sigma}_{zy})}{tr(Q_{yy} \hat{\Sigma}_{yy})}.$

Once the partial derivatives of ρ_{sur}^2 with respect to $(vec\Sigma)'$ are obtained, the limiting distribution of $\hat{\rho}_{sur}^2$ directly follows from the asymptotic normality of (vec$\hat{\Sigma}$). Naturally one can use the derivatives that are already obtained in the preceding section for the p=m version of the SE coefficient of linear association as in this case ρ_{se}^2 and ρ_{sur}^2 coincide (see (3.27)). However, in this formula the characteristic roots and vectors of $(\Sigma_{yy}-\Sigma'_{zy}\Sigma_{zz}^{-1}\Sigma_{zy})$ are employed, while they are of no significance for the determination of the SUR parameters. Therefore, in the following theorem the corresponding results will be given, not by using these SE characteristics, but by means of the SUR parameter matrix B_{sur}.

The matrix of derivatives for p=1 are already given by Van Praag (1980) and Van Praag, Dijkstra, Van Velzen (1985). In the latter paper it was argued that, in order to apply the delta method, it should be assumed that $0 < \rho_{sur}^2 < 1$, which is in our case implied by the fact that Σ (and therefore also Σ_{yy}, $\Sigma'_{zy}\Sigma_{zz}^{-1}\Sigma_{zy}$ and $\Sigma_{yy}-\Sigma'_{zy}\Sigma_{zz}^{-1}\Sigma_{zy}$) is a positive definite matrix. The coefficient of linear association is in these papers observed as a function of the $(((k-1)^2+k)\times1)$ vector $((vec\Sigma_{zz})', \Sigma_{zy}, \Sigma_{yy})'$ and therefore the partial derivatives of ρ_{sur}^2 (for p=1) were derived with respect to $(vec\Sigma_{zz})'$, Σ'_{zy} and Σ_{yy}. In the following theorem corresponding results are given in the more general setting where $p \geq 1$, with respect to the vectorized complete covariance matrix $(vec\Sigma)'$.

<u>Theorem</u> (the derivative of the SUR coefficient of linear association)
Let ρ_{sur}^2 be defined as indicated in (4.3), then ρ_{sur}^2 may be considered as a differentiable function of Σ with

$$(4.8) \qquad \nabla(\rho_{sur}^2;\Sigma) = \frac{\left(vec \left[\begin{array}{c|c} -\rho_{sur}^2 Q_{yy} & 0 \\ \hline \dfrac{1}{2B}_{sur,z}Q_{yy} & -B_{sur,z}Q_{yy}B'_{sur,z} \end{array}\right]\right)'}{tr(Q_{yy}\Sigma_{yy})} \qquad (1\times k^2)$$

with $B_{sur,z}$ the SUR parameter matrix given in (4.2).

Prior to the proof of this proposition, the resulting gradient vector is compared with the corresponding vector of partial derivatives in the context of SE analysis as denoted in (3.23). Rewriting (4.8) as

$$(4.9) \qquad \nabla(\rho^2_{sur};\Sigma) = \frac{(vec((1-\rho^2_{sur})Q_o - B_{sur}Q_{yy}B'_{sur} - \left[\begin{array}{c|c} 0 & Q_{yy}B'_{sur,z} \\ \hline -B_{sur,z}Q_{yy} & 0 \end{array}\right]))'}{tr(Q_{yy}\Sigma_{yy})}$$

it appears that the dissimilarities in the formula notation, caused by the fact that not B_{se}, but B_{sur} is used, are in this last term of the gradient vector. Furthermore, when p=1 the matrix B_{sur} degenerates into a column vector β_{sur} and (4.8) can be rewritten as

$$(4.10) \qquad \nabla(\rho^2_{sur};\Sigma) = \frac{1}{\Sigma_{yy}}\left[-\rho^2_{sur} \mid 2\beta'_{sur,z} \mid -\beta'_{sur,z} \otimes \left[0 \mid \beta'_{sur,z}\right]\right]'$$

which corresponds with the results obtained in the aforementioned papers.

Proof

As ρ^2_{sur} denotes a quotient of two scalar functions in Σ, the quotient rule for differentiation can be applied and the row vectors of partial derivatives is given by

$$(4.11) \qquad \nabla(\rho^2_{sur};\Sigma) = \frac{tr(Q_{yy}\Sigma_{yy}) \ \nabla(tr(Q_{yy}\Sigma'_{zy}\Sigma^{-1}_{zz}\Sigma_{zy});\Sigma)}{(tr(Q_{yy}\Sigma_{yy}))^2}$$

$$- \frac{tr(Q_{yy}\Sigma'_{zy}\Sigma^{-1}_{zz}\Sigma_{zy}) \ \nabla(tr(Q_{yy}\Sigma_{yy});\Sigma)}{tr(Q_{yy}\Sigma_{yy}))^2}$$

$$= \frac{\nabla(tr(Q_{yy}\Sigma'_{zy}\Sigma^{-1}_{zz}\Sigma_{zy});\Sigma) - \rho^2_{sur} \ \nabla(tr(Q_{yy}\Sigma_{yy});\Sigma)}{tr(Q_{yy}\Sigma_{yy})}.$$

Hence $\nabla(\rho^2_{sur};\Sigma)$ is rewritten as a function of the derivatives of the numerator and the denominator of ρ^2_{sur}. Let us first observe the numerator. Applying (A.14), the trace is rewritten as a product of two vectors, yielding

$$(4.12) \qquad \nabla(tr(Q_{yy}\Sigma'_{zy}\Sigma^{-1}_{zz}\Sigma_{zy});\Sigma) = \nabla((vec(\Sigma_{zy}Q_{yy}))'(vec(\Sigma^{-1}_{zz}\Sigma_{zy}));\Sigma)$$

which is by means of the product rule (A.41) rewritten as

(4.13) $\quad \dfrac{\partial(\text{vec}(\Sigma_{zy}Q_{yy}))'}{\partial(\text{vec}\Sigma)'}\ (\text{vecB}_{\text{sur},z}\ \otimes\ I_{k^2}) + (\text{vec}(\Sigma_{zy}Q_{yy}))'\ \nabla(B_{\text{sur},z};\Sigma).$

The second factor of the last term is known from section III.4 (see (III.4.14)) and is replaced by its outcome. Rewriting Σ_{zy} in the first term by $[0\ \vdots\ I_{k-p}]\Sigma[I_p\ \vdots\ 0]'$, (4.13) is rewritten as

(4.14) $\quad \dfrac{\partial(\text{vec}([0\ \vdots\ I_{k-p}]\Sigma[Q_{yy}\ \vdots\ 0]'))'}{\partial(\text{vec}\Sigma)'}\ (\text{vecB}_{\text{sur},z}\ \otimes\ I_{k^2})$

$\qquad - (\text{vec}(\Sigma_{zy}Q_{yy}))'(B'_{\text{sur}}\ \otimes\ \begin{bmatrix} 0\ \vdots\ \Sigma_{zz}^{-1} \end{bmatrix}).$

According to (A.11), $\text{vec}([0\ \vdots\ I_{k-p}]\Sigma[Q_{yy}\ \vdots\ 0]')$ can be replaced by $([Q_{yy}\ \vdots\ 0]\ \otimes\ [0\ \vdots\ I_{k-p}])(\text{vec}\Sigma)$. Using the matrix product differentiation rule (A.41) in relation with the resulting first term of (4.14) we obtain

(4.15) $\quad \dfrac{\partial(\text{vec}\Sigma)'}{\partial(\text{vec}\Sigma)'}\ (([Q_{yy}\ \vdots\ 0]'\ \otimes\ [0\ \vdots\ I_{k-p}]')\ \otimes\ I_{k^2})(\text{vecB}_{\text{sur},z}\ \otimes\ I_{k^2})$

$\qquad - (\text{vec}(\Sigma_{zy}Q_{yy}))'(B'_{\text{sur}}\ \otimes\ \begin{bmatrix} 0\ \vdots\ \Sigma_{zz}^{-1} \end{bmatrix}).$

Observing that the result of $\dfrac{\partial(\text{vec}\Sigma)'}{\partial(\text{vec}\Sigma)'}$ is given by $(\text{vecI}_{k^2})'$ and again applying (A.11) to both the first and second term, (4.15) can be rewritten as

(4.16) $\quad (\text{vecI}_{k^2})(\text{vec}([0\vdots B'_{\text{sur},z}]'[Q_{yy}\vdots 0])\ \otimes\ I_{k^2}) - (\text{vec}(\begin{bmatrix} 0\vdots\Sigma_{zz}^{-1} \end{bmatrix}'\Sigma_{zy}Q_{yy}B'_{\text{zur}}))'$

which equals

(4.17) $\quad (\text{vec}\begin{bmatrix} 0 & \vdots & 0 \\ B_{\text{sur},z}Q_{yy} & \vdots & 0 \end{bmatrix})' - (\text{vec}\begin{bmatrix} 0 & \vdots & 0 \\ -B_{\text{sur},z}Q_{yy} & \vdots & B_{\text{sur},z}Q_{yy}B'_{\text{sur},z} \end{bmatrix})'.$

Hence the first term of the numerator of (4.11) is given by

$$(4.18) \qquad \nabla(\text{tr}(Q_{yy}\Sigma'_{zy}\Sigma^{-1}_{zz}\Sigma_{zy});\Sigma) = (\text{vec}\left[\begin{array}{c|c} 0 & 0 \\ \hline 2B_{sur,z}Q_{yy} & -B_{sur,z}Q_{yy}B_{sur,z} \end{array}\right])'.$$

Then we come to the derivative of the denominator of ρ^2_{sur}, which is defined by

$$(4.19) \qquad \nabla(\text{tr}(Q_{yy}\Sigma_{yy});\Sigma) = (\text{vec}\ \frac{\partial(\text{tr}(Q_{yy}\Sigma_{yy}))}{\partial\Sigma})'$$

$$= (\text{vec}\ \frac{\partial(\text{tr}([Q_{yy} \vdots 0]\Sigma[\Sigma_p \vdots 0]'))}{\partial\Sigma})'$$

Application of (A.49) yields

$$(4.20) \qquad \nabla(\text{tr}(Q_{yy}\Sigma_{yy});\Sigma) = (\text{vec}\left[\begin{array}{c|c} Q_{yy} & 0 \\ \hline 0 & 0 \end{array}\right])'.$$

Combining (4.11) together with (4.18) and (4.20) we obtain the expression for $\nabla(\rho^2_{sur};\Sigma)$ as proposed in (4.8). ◇

As a result of this theorem, we can derive the asymptotic variance of $\hat{\rho}^2_{sur}$. Denoting by V the covariance matrix of the limiting distribution of $\text{vec}\Sigma$, obtained from a sample consisting of N observations, the limiting distribution of $\hat{\rho}^2_{sur}$ is given by

$$(4.21) \qquad \sqrt{N}(\hat{\rho}^2_{sur} - \rho^2_{sur}) \xrightarrow{D} N(0,\ [\nabla(\rho^2_{sur};\Sigma)]V[\nabla(\rho^2_{sur};\Sigma)]'))$$

with $\nabla(\rho^2_{sur};\Sigma)$ given in (4.8). When moreover the observations are i.i.d., all with the same distribution as X, then V stands for the variance matrix of $\text{vec}(XX')$ and a simplified expression for the asymptotic variance can be obtained as follows

Theorem ($\hat{\rho}^2_{sur}$ for random i.i.d. sample observations)
Let $\{X_n: n=1,\ldots,N\}$ be a sample of N i.i.d. random observations on X and let $\hat{\rho}^2_{sur}$ be the MME of ρ^2_{sur}, that is defined as the coefficient of linear association in the SUR analysis (see (4.3)). Then

$$(4.22) \quad \sqrt{N}(\hat{\rho}^2_{sur} - \rho^2_{sur}) \xrightarrow{D} \mathbb{N}(0, \frac{E[(1-\rho^2_{sur})Y'Q_{yy}Y - (Y-B'_{sur,z}Z)'(Y-B'_{sur,z}Z)]^2}{(tr(Q_{yy}\Sigma_{yy}))^2}).$$

Proof

In order to prove (4.22) we only need to derive the asymptotic variance by pre- and post-multiplication of the covariance matrix of vec(XX') by $\nabla(\rho^2_{sur};\Sigma)$ and its transpose. Substitution of (4.8) for $\nabla(\rho^2_{sur};\Sigma)$, replacing V by $E(vec(XX'))(vec(XX'))' - (vec\Sigma)(vec\Sigma)'$ and using the partitioning of X in Y and Z we obtain by using (A.14)

$$(4.23) \quad [\nabla(\rho^2_{sur};\Sigma)]V[\nabla(\rho^2_{sur};\Sigma)]' =$$

$$= \frac{E[tr(Q_{yy}(-\rho^2_{sur}YY' + 2B'_{sur,z}ZY' - B'_{sur,z}ZZ'))]^2}{(tr(Q_{yy}\Sigma_{yy}))^2}$$

$$- \frac{[tr(Q_{yy}(-\rho^2_{sur}\Sigma_{yy} + 2B'_{sur,z}\Sigma_{zy} - B'_{sur,z}\Sigma_{zz}))]^2}{(tr(Q_{yy}\Sigma_{yy}))^2}.$$

Replacing in the argument of the last trace operator $B_{sur,z}$ by $\Sigma_{zz}^{-1}\Sigma_{zy}$, it can be shown that it equals the zero matrix, and the remaining non-zero part at the right-hand side of (4.23) can be rewritten as

$$(4.24) \quad \frac{E[tr(Q_{yy}(1-\rho^2_{sur})YY' - (Y-B'_{sur,z}Z)(Y-B'_{sur,z}Z)')]^2}{tr(Q_{yy}\Sigma_{yy}))^2}$$

from which the asymptotic variance as proposed in (4.22) follows directly. ◊

It has already been noted that in case p=1, $\hat{\rho}^2_{sur}$ equals the multiple correlation coefficient R^2. When moreover k=2, then $\hat{\rho}^2_{sur}$ equals the correlation coefficient between two sets of random variables, which has in (1.8) been denoted by r^2_{yz}. The derivation of the (asymptotic) distributional properties of these kind of statistics has a long history. The distribution of r^2_{yz} under the assumption of bivariate, normally distributed observations for which $cov(y_n,z_n) = 0$, was derived by Fisher (1915). He also proposed a monotonic transformation $\frac{1}{2}\ln((1+r_{yz}^2)/(1-r_{yz}^2))$ in case $cov(y_n, z_n) \neq 0$, which produces an asymptotically normal variate, usually denoted by z (Fisher's z-transformation (1921)). The distribution of R^2 computed from multinormal distributed observations was derived by Fisher (1928) and Wishart (1931). Muirhead (1980) derived some asymptotic results for the multiple correlation coefficients subject to the assumption of elliptically distributed sample observations.

So far, all the coefficients of linear association we have dealt with can be interpreted as the squared cosine of an angle. This was caused by the fact that $\tilde{X}$ is obtained from X by an orthogonal projection. This will generally not be the case when we place some restrictions on the coefficients of the exogenous variables. As an example the goodness-of-fit measure as denoted in (4.3) will be observed in the restricted seemingly unrelated regression (RSUR) analysis. In RSUR analysis we are also interested in the relation as denoted in (4.1), that is in $\tilde{Y} = B_z'Z$, however, now the optimal value of B_z is determined subject to the restriction $R'(vecB_z) = c$, with R a $((k-p)p\times r)$ matrix and c an $(r\times 1)$ vector of constants. The solutions have been derived in section III.5 and are given by (see (III.5.5) and (III.5.6))

$$(4.25) \qquad vecB_{rsur,z} = vecB_{sur,z} + (Q_{yy}^{-1} \otimes \Sigma_{zz}^{-1})R\kappa_o$$

with

$$(4.26) \qquad \kappa_o = (R'(Q_{yy}^{-1} \otimes \Sigma_{zz}^{-1})R)^{-1}(c-R'(vecB_{sur,z})).$$

Evaluating $\rho_{Q_{yy}}^2$ (which is defined in (3.1)) in $\tilde{Y} = B_{rsur,z}'Z$, we obtain an expression for the coefficients of linear association, say ρ_{rsur}^2, in the RSUR case. The three factors in ρ_{rsur}^2 are given by

$$(4.27) \qquad E(Y'Q_{yy}Y) = tr(Q_{yy}\Sigma_{yy})$$

$$(4.28) \qquad E(Y'Q_{yy}\tilde{Y}) = tr(Q_{yy}B_{rsur,z}'\Sigma_{zy})$$

and

$$(4.29) \qquad E(\tilde{Y}'Q_{yy}\tilde{Y}) = tr(Q_{yy}B_{rsur,z}'\Sigma_{zz}B_{rsur,z}).$$

Using (A.14) and substituting (4.25) for $vecB_{rsur,z}$, (4.29) can be rewritten as

$$(4.30) \qquad E(\tilde{Y}'Q_{yy}\tilde{Y}) = (vec(\Sigma_{zz}B_{rsur,z}Q_{yy}))'(vecB_{sur,z} + (Q_{yy}^{-1} \otimes \Sigma_{zz}^{-1})R\kappa_o).$$

Application of (A.14) and (A.11) to the first and second term of (4.30) respectively yields

$$(4.31) \qquad E(\tilde{Y}'Q_{yy}\tilde{Y}) = tr(Q_{yy}B'_{rsur,z}\Sigma_{zy}) + (vecB_{rsur,z})'R\kappa_o$$

$$= tr(Q_{yy}B'_{rsur,z}\Sigma_{zy}) + c'\kappa_o.$$

Combining (4.27), (4.28) and (4.31) we obtain

$$(4.32) \qquad \rho^2_{rsur} = \frac{(tr(Q_{yy}B'_{rsur,z}\Sigma_{zy}))^2}{tr(Q_{yy}\Sigma_{yy})(tr(Q_{yy}B'_{rsur,z}\Sigma_{zy}) + c'\kappa_o)}.$$

Note that, although B_z is evaluated in its optimum, this does not yield the equality between $E(Y'Q_{yy}\tilde{Y})$ and $E(\tilde{Y}'Q_{yy}\tilde{Y})$. Only in case of a homogeneous restriction on B_z (that is when c=0), the coefficient of linear association can be simplified as

$$(4.33) \qquad \rho^2_{rsur} = \frac{tr(Q_{yy}B'_{rsur}\Sigma_{zy})}{tr(Q_{yy}\Sigma_{yy})}$$

which can be shown to yield the implication

$$(4.34) \qquad c = 0 \Rightarrow \rho^2_{rsur} = \rho^2_{sur} + \frac{\kappa_o'R'(vecB_{sur,z})}{tr(Q_{yy}\Sigma_{yy})}.$$

Note that the vector κ_o of Lagrange multipliers indicates for both the parameter matrix $B_{rsur,z}$ and the coefficient ρ^2_{rsur}, to what extent they diverge from their unrestricted variants. Hence the Lagrange multipliers measure the "goodness" of the prior information, here given by the restriction

$R'(vecB_z) = 0$. Estimators of κ_0 can be used to test whether the hypothesized linear model fulfills the prior information. A so-called Lagrangian multiplier test has been developed by Aitchinson and Silvey (1958, 1960) and Silvey (1959) in the context of restricted Maximum Likelihood estimation. Applications of these tests to parametric model specifications have for instance been given by Byron (1970) and Breusch and Pagan (1980). The estimator of ρ^2_{sur} measures to what extent the hypothesized linear model fulfills the sample information. From (4.36) it follows that ρ^2_{rsur} measures the goodness-of-fit caused by a mixture of sample- and non-random prior information. Clearly the values of κ_0 and ρ^2_{sur} are of more importance than the value of ρ^2_{rsur}, as they measure the influence of the sample and the restriction on the parameter estimators separately. Therefore we shall not elaborate on the derivatives of ρ^2_{rsur} and the limiting distribution of its moments estimators, although this would not be very difficult as the formulas follow almost directly from the corresponding results for $B_{rsur,z}$ and ρ^2_{sur}.

VI. Review

VI.1 A schematic representation of the parameters

Although this study has not yet been completed, the author believes, that a
review of the preceding results is appropriate at this point. So far, the
treatise on the statistical problem of searching for linear relations in
observations has been purely theoretical. In chapters III, IV and V parameter
matrices have been obtained as the optimal values of distance minimizing
problems. They are observed as (explicitly or implicitly defined) functions of
the population covariance matrix Σ. In chapter II an explanation has been
given of how asymptotically normally distributed estimators of this class of
parameters can be obtained, together with their large sample properties, given
that the parameters in question are differentiable in Σ. It appeared that in
order to apply the delta-method, the derivatives of the parameter matrices
with respect to the elements of Σ are needed. Therefore each section that
discusses a distance minimizing parameter matrix B (or A), also provides the
corresponding gradient matrix $\nabla(B;\Sigma)$ (or $\nabla(A;\Sigma)$) of the partial derivatives.
In chapter VII the Population-Sample Decomposition approach will be used to
show how these distance minimizing parameters can be applied in practice, and
it will be made clear which problems have to be overcome, in order to progam
the PSD estimation technique efficiently.

This chapter does not only serve as a buffer between the theoretical and the
practical parts of this study, but it also tries to throw some light on the
systematic construction of the preceding theoretical chapters. This will be
done in the present section (VI.1) on the basis of figure VI.1.1. We have seen
in chapters III, IV and V, where each section dealt with one parameter matrix
and its gradient matrix, that the results are formulated in theorems. As the
reader is in danger of getting lost among the large number of formulas used in
the proofs of the theorems, it seems wise to recapitulate the principal
formulas in a concise summary. This will be done in section VI.2.

Let us first review the classification of the parameter matrices that have
been considered in the various chapters and sections of the study. To do this,
it is necessary to turn to the schematic representation of the sections, given
in figure VI.1.1. Here we find the organization of the sections of the
preceding four chapters in the form of a tree diagram. The title of

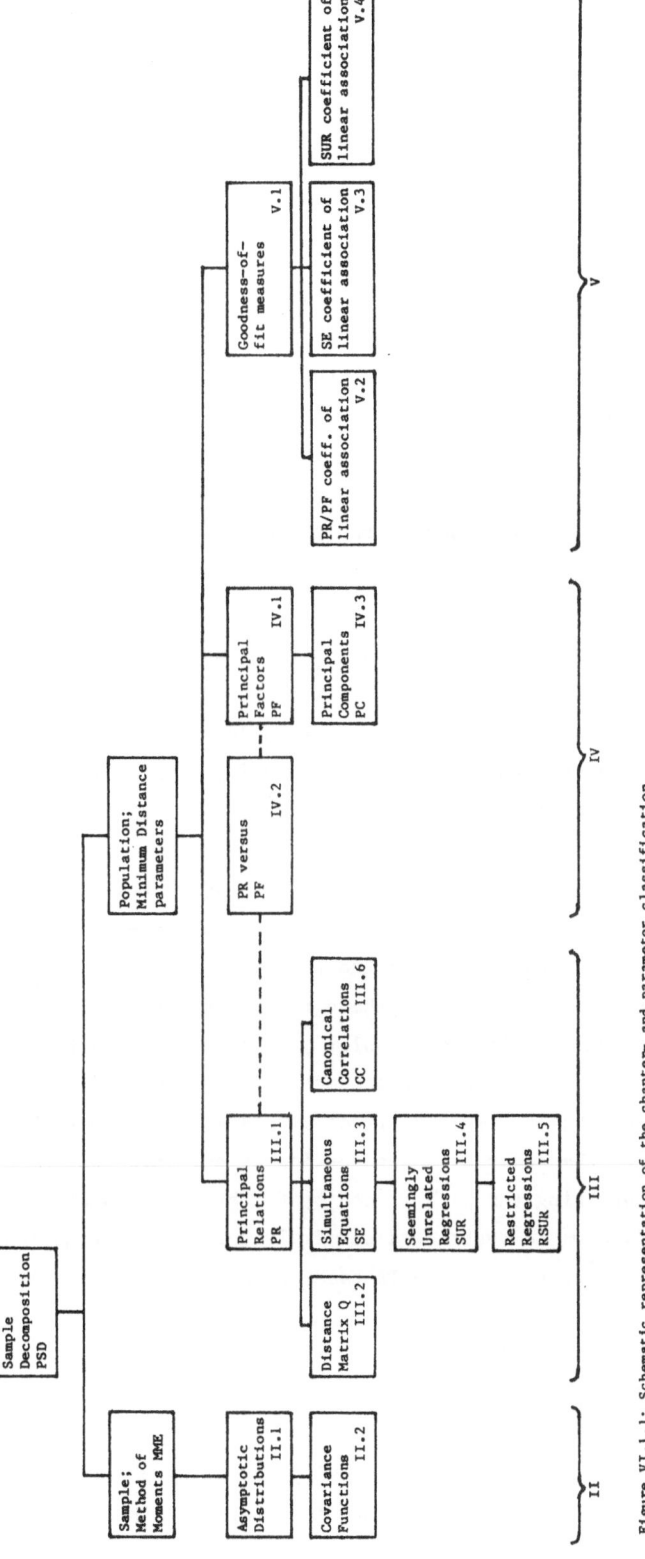

Figure VI.1.1: Schematic representation of the chapter- and parameter classification

this study is to be found in the root of the tree diagram; it refers to both
the population part and the sample part of the PSD estimation technique. The
two branches which lead to the second level of the tree represent the word
"decomposition" which is to be found it the title: the left branch connects up
with the sample part, whilst the right branch connects up with the population
part. The name of the estimation technique applied in this study, viz "Method
of Moments", and the name of the class of parameters, viz. "Minimum Distance
parameters", make up the contents of the two second level nodes. The remaining
nodes, that make up the third to the sixth level of the tree diagram, contain
the following information. In the centre of the node we find the name of a
statistical linear analysis method. This method can be considered as a special
case of the technique recorded in the preceding node. In the bottom right-hand
corner of each node, the number of the section is recorded in which the
associated parameters are dealt with and the relevant abbreviation (if
existent) is given in the bottom left-hand corner. The graph is organized in
such a way that the nodes corresponding with the sections of one chapter are
to be found all together in one cluster. These nodes are joined up in clusters
by means of wavy brackets and the corresponding chapter numbers are given in
Roman numerals at the bottom of the figure.

Let us confine ourselves to chapters III, IV and V. What is the idea behind
the organization of the different sections in this tree diagram? At the third
level of the tree diagram we find the nodes that refer to the first sections
of each chapter. Recall, that in these first sections the basic formulations
were given of the minimization problems, to be studied in that chapter. By
defining specific structures for the metric matrix Q and/or placing
restrictions on the parameter matrices, we isolate a number of special cases
appertaining to the analysis under consideration. This concept will be made
clear by illustrating the path traced by the branches running from principal
relations (PR) down to restricted seemingly unrelated regressions (RSUR).
The minimization problem in the PR case is formulated as

$$(1.1) \qquad \min_{\substack{B \\ B'\xi=0}} E[\ \min_{\xi}\ (X-\xi)'Q(X-\xi)].$$

For an explanation of this mathematical programming problem and the dimensions
of the vectors and matrices used, one is referred to section III.1. The

solution yields the optimal B matrix in the PR analysis, denoted by B_{pr}. Then we start to distinguish two parts of vector X, denoted by Y and Z.

The vector ξ and matrices B and Q are partitioned accordingly, yielding

$$(1.2) \qquad \xi = \begin{bmatrix} \eta \\ \zeta \end{bmatrix}, \quad B = \begin{bmatrix} B_y \\ \hline B_z \end{bmatrix} \text{ and } Q = \begin{bmatrix} Q_{yy} & Q'_{zy} \\ \hline Q_{zy} & Q_{zz} \end{bmatrix}.$$

In section III.3 the mathematical programming problem (1.1) has been observed for a special class of metric matrices, viz.

$$(1.3) \qquad Q = \lim_{q \to \infty} \begin{bmatrix} Q_{yy} & 0 \\ \hline 0 & q \ddots \\ & & q \end{bmatrix}.$$

Hence only the submatrix Q_{yy} can be freely chosen. The type of relations in X, generated by this optimalization problem, is called the simultaneous equations (SE) structure, as it manifests various correspondences with the statistical technique of the same name, that is widely used in econometric theory. The optimal value of B is denoted by B_{se}. In section III.4 the minimization problem (1.1) with the special class of metric matrices is considered together with the restriction that B_y equals $-I_p$. This means that only the optimal value of B_z has to be derived, which is denoted by $B_{sur,z}$, where SUR stands for the abbreviation of seemingly unrelated regressions. Finally, not only do we restrict B_y to equal $-I_p$, but we also place linear restrictions upon the elements of B_z. These restrictions are indicated by the following formula

$$(1.4) \qquad R'(\text{vec}B_z) = c,$$

with R and c being respectively a matrix and a vector of a priori known constants, yielding a number of equality constraints for B_z. The conclusion is that each node of this path denotes a population technique, which equals the analysis described in the preceding node, together with an additional assumption about the form of Q or B. The reader is encouraged to check the remaining branches of the tree in a similar way.

VI.2 List of notation and summary of results

This section is meant to be a concise review of the formulas obtained in chapters III, IV and V. These formulas have been derived in order to apply the so-called Population-Sample Decomposition (PSD) estimation technique, which was introduced in chapter II. The population parameters of interest have been derived as (matrix-valued) functions of Σ, which are denoted by B, A and ρ^2 (or more formally by functions $B(\Sigma)$, $A(\Sigma)$ and $\rho^2(\Sigma)$) respectively. Each section of chapters III and IV (except for III.2 and IV.2) was built up of three segments. These segments are briefly described as follows:

(i)   Definition of a distance minimization problem.
(ii)   The solution of the minimum distance problem, yielding a parameter matrix, defined as a (not necessarily explicit) function, differentiable in Σ, denoted by $B_.$ or $A_.$.
(iii)   The matrix of partial derivatives of the optimal parameter matrix, evaluated in Σ, denoted by $\nabla(B_.;\Sigma)$ or $\nabla(A_.;\Sigma)$.

In the summary of the results we shall list the main formulas in the order given above. The train of thought behind the design of the minimization problem has been dealt with extensively in the preceding sections and will not be considered here. In chapter V, which dealt with the goodness-of-fit measures of the aforementioned linear structural models, the coefficients were not obtained as the optimal values of a minimization problem. However, once again the parameters of interest are obtained as functions that are differentiable in Σ. Therefore it is useful to review the properties of the goodness-of-fit measures according to the following scheme:

(a) Definition of the goodness-of-fit measure as a function, differentiable in Σ, denoted by $\rho^2_.$.
(b) The matrix of partial derivatives of $\rho^2_.$, evaluated in Σ, denoted by $\nabla(\rho^2_.;\Sigma)$.

Before proceeding to the summary of results, a list of notations for the vectors and matrices that have been used in the chapters III, IV and V will be presented. Matrices and vectors are presented together with their dimensions and a short description. For the completeness of the summary even the most elementary concepts will be repeated, to ensure that this review will be self-contained and that no references to preceding formulas are needed.

List of notation

X	$(k \times 1)$ random vector
ξ	$(k \times 1)$ auxiliary vector
υ	$(q \times 1)$ auxiliary vector
$\Sigma = E(XX')$	$(k \times k)$ covariance matrix of X
$B_. = [\beta_{.,1}, \ldots, \beta_{.,p}]$	$(k \times p)$ parameter matrix of rank p, with $p < k$
$A_. = [\alpha_{.,1}, \ldots, \alpha_{.,q}]$	$(k \times q)$ parameter matrix of rank q, with $q < k$
Q	$(k \times k)$ symmetric, positive definite matrix
$Q^{\frac{1}{2}}$	$(k \times k)$ symmetric matrix, such that $Q^{\frac{1}{2}} Q^{\frac{1}{2}} = Q$
$\Sigma^* = Q^{\frac{1}{2}} \Sigma Q^{\frac{1}{2}}$	$(k \times k)$ transformed covariance matrix
$\lambda_1 > \ldots > \lambda_k$	the characteristic roots of Σ in the Euclidean metric defined by Q^{-1}
$\Lambda_k = \text{diag}(\lambda_1, \ldots, \lambda_k)$	$(k \times k)$ diagonal matrix
$\Lambda_{\min,p} = \text{diag}(\lambda_k, \ldots, \lambda_{k-p+1})$	$(p \times p)$ diagonal matrix
$\Lambda_{\max,q} = \text{diag}(\lambda_1, \ldots, \lambda_q)$	$(q \times q)$ diagonal matrix
$\lambda_{pc,1} > \ldots > \lambda_{pc,k}$	the characteristic roots of Σ in the Euclidean metric defined by I_k
$\Lambda_{pc,\max,q} = \text{diag}(\lambda_{pc,1}, \ldots, \lambda_{pc,q})$	$(q \times q)$ diagonal matrix
R	$((k-p)p \times r)$ matrix of constants of rank r, with $r \leq (k-p)p$
c	$(r \times 1)$ vector of constants
$X = \begin{bmatrix} Y \\ Z \end{bmatrix}$	$(m \times 1)$ vector Y and $((k-m) \times 1)$ vector Z, with $p \leq m < k$

$$\xi = \begin{bmatrix} \eta \\ \zeta \end{bmatrix}$$

$$\Sigma = \begin{bmatrix} \Sigma_{yy} & \vdots & \Sigma'_{zy} \\ \hline \Sigma_{zy} & \vdots & \Sigma_{zz} \end{bmatrix}$$

$$B_{\cdot} = \begin{bmatrix} B_{\cdot,y} \\ \hline B_{\cdot,z} \end{bmatrix}$$

Partitioning of ξ, Σ, $B_{\cdot}$ and Q conformably with X

$$Q = \begin{bmatrix} Q_{yy} & \vdots & Q'_{zy} \\ \hline Q_{zy} & \vdots & Q_{zz} \end{bmatrix}$$

$$Q_{cc} = \begin{bmatrix} \Sigma_{yy}^{-1} & \vdots & 0 \\ \hline 0 & \vdots & \Sigma_{zz}^{-1} \end{bmatrix}$$

specific value of matrix Q

$Q_{cc}^{\frac{1}{2}}$

($k \times k$) symmetric matrix, such that $Q_{cc}^{\frac{1}{2}} Q_{cc}^{\frac{1}{2}} = Q_{cc}$

$\Sigma_{cc}^{*} = Q_{cc}^{\frac{1}{2}} \Sigma Q_{cc}^{\frac{1}{2}}$

($k \times k$) transformed covariance matrix

$\lambda_{cc,1} > \cdots > \lambda_{cc,k}$

the characteristic roots of Σ in the Euclidean metric defined by Q_{cc}^{-1}

$\phi_i = 1 - \lambda_{cc,k-i+1}$

canonical correlations for $i=1,\ldots,p$

$\Phi_{max,p} = diag(\phi_1, \ldots, \phi_p)$

($p \times p$) diagonal matrix

$Q_{yy}^{\frac{1}{2}}$

($m \times m$) symmetric matrix, such that $Q_{yy}^{\frac{1}{2}} Q_{yy}^{\frac{1}{2}} = \Sigma_{yy}$

$\Sigma_{yy}^{*} = Q_{yy}^{\frac{1}{2}} \Sigma_{yy} Q_{yy}^{\frac{1}{2}}$

($m \times m$) transformed covariance sub-matrix

$\Sigma_{zy}^{*} = \Sigma_{zy} Q_{yy}^{\frac{1}{2}}$

(($k-m) \times m$) transformed covariance sub-matrix

$\omega_1 > \cdots > \omega_m$

the characteristic roots of $(\Sigma_{yy} - \Sigma'_{zy} \Sigma_{zz}^{-1} \Sigma_{zy})$ in the metric defined by Q_{yy}^{-1}

$\Omega_{min,p} = diag(\omega_m, \ldots, \omega_{m-p+1})$

($p \times p$) diagonal matrix

A list of the main formulas for the parameter matrices and the goodness-of-fit coefficients, according to schemes (i), (ii), (iii) and (a), (b) respectively, is given in the following review.

Summary of results

Principal Relations (PR):

(i) $\min\limits_{B} E[\min\limits_{\substack{\xi \\ B'\xi=0}} (X-\xi)'Q(X-\xi)]$

(ii) $\Sigma B_{pr} = Q^{-1}B_{pr}\Lambda_{min,p}$ and $B'_{pr}Q^{-1}B_{pr} = I_p$

(iii)   for i=1,...,p:

$$V(\beta_{pr,i};\Sigma) = -(\beta'_{pr,i} \otimes Q^{\frac{1}{2}}(\Sigma^* - \lambda_{k-i+1}I_k)^+Q^{\frac{1}{2}})$$

Simultaneous Equations (SE):

(i) $\min\limits_{B} E[\min\limits_{\substack{\eta \\ B'_y\eta + B'_z Z=0}} (Y-\eta)'Q_{yy}(Y-\eta)]$

(ii) $(\Sigma_{yy} - \Sigma'_{zy}\Sigma_{zz}^{-1}\Sigma_{zy})B_{se,y} = Q_{yy}^{-1}B_{se,y}\Omega_{min,p}$ and $B_{se,y}Q_{yy}^{-1}B_{se,y} = I_p$

$B_{se,z} = -\Sigma_{zz}^{-1}\Sigma_{zy}B_{se,y}$

(iii)   for i=1,...,p:

$$V(\beta_{se,y,i};\Sigma) = -(\beta'_{se,i} \otimes \left[Q_{yy}^{\frac{1}{2}}((\Sigma_{yy}^* - \Sigma_{zy}^{*'}\Sigma_{zz}^{-1}\Sigma_{zy}^*) - \omega_{m-i+1}I_m)^+Q_{yy}^{\frac{1}{2}} \mid 0\right])$$

$$V(\beta_{se,z,i};\Sigma) = -\Sigma_{zz}^{-1}\Sigma_{zy} V(\beta_{se,y,i};\Sigma) - (\beta'_{se,i} \otimes \left[0 \mid \Sigma_{zz}^{-1}\right])$$

Seemingly Unrelated Regressions (SUR):

(i) $\min\limits_{B_z} E[(Y-B'_z Z)'Q_{yy}(Y-B'_z Z)]$

(11) $B_{sur,y} = -I_p$ and $B_{sur,z} = \Sigma_{zz}^{-1}\Sigma_{zy}$

(iii) $\nabla(B_{sur,z};\Sigma) = -(B'_{sur} \otimes \left[\, 0 \;\vdots\; \Sigma_{zz}^{-1}\right])$

Restricted Seemingly Unrelated Regressions (RSUR):

(i) $\min_{B_z} E[(Y-B'_z Z)'Q_{yy}(Y-B'_z Z)]$

         $R'(vecB_z) = c$

(ii) $B_{rsur,y} = -I_p$

         $B_{rsur,z} = B_{sur,z} + (Q_{yy}^{-1} \otimes \Sigma_{zz}^{-1})R(R'(Q_{yy}^{-1} \otimes \Sigma_{zz}^{-1})R)^{-1}(c-R'(vecB_{sur,z}))$

(iii) $\nabla(B_{rsur,z};\Sigma) = -[I_{(k-p)p} - (Q_{yy}^{-1} \otimes \Sigma_{zz}^{-1})R(R'(Q_{yy}^{-1} \otimes \Sigma_{zz}^{-1})R)^{-1}R']$

         $\times (B'_{rsur} \otimes \left[\, 0 \;\vdots\; \Sigma_{zz}^{-1}\right])$

Canonical Correlations (CC):

(i) $\min_B E[\min_{\substack{\xi \\ B'\xi=0}} (X-\xi)'Q_{cc}(X-\xi)]$

(ii) $\Sigma'_{zy}\Sigma_{zz}^{-1}\Sigma_{zy}B_{cc,y} = \Sigma_{yy}B_{cc,y}\phi^2_{max,p}$, $\Sigma_{zy}\Sigma_{yy}^{-1}\Sigma'_{zy}B_{cc,z} = \Sigma_{zz}B_{cc,z}\phi^2_{max,p}$,

         $B'_{cc,y}\Sigma_{yy}B_{cc,y} = I_p$ and $B'_{cc,z}\Sigma_{zz}B_{cc,z} = I_p$

(iii)   for i=1,...,p:

$$\nabla(\beta_{cc,i};\Sigma) = -(\beta'_{cc,i} \otimes Q_{cc}^{\frac{1}{2}}(\Sigma^*_{cc}-\lambda_{cc,k+i-1}I_k)^+Q_{cc}^{\frac{1}{2}})$$

$$+ \lambda_{cc,k+i-1}Q_{cc}^{\frac{1}{2}}(\Sigma^*-\lambda_{cc,i}I_k)^+Q_{cc}^{\frac{1}{2}}$$

$$\times \left[\beta'_{cc,y,i} \otimes \begin{bmatrix} I_m & \vdots & 0 \\ \hline 0 & \vdots & 0 \end{bmatrix} \vdots \beta'_{cc,z,i} \otimes \begin{bmatrix} 0 & \vdots & 0 \\ \hline 0 & \vdots & I_{k-m} \end{bmatrix}\right]$$

Principal Factors (PF):

(i) $\min_{A} E[\min_{\upsilon} (X-A\upsilon)'Q(X-A\upsilon)]$

(ii) $\Sigma Q A_{pf} = A_{pf}\Lambda_{max,q}$ and $A'_{pf}Q A_{pf} = \Lambda_{max,q}$

(iii)   for i = 1,...,q:

$$\nabla(\alpha_{pf,i};\Sigma) = -Q^{-1}(\alpha'_{pf,i}Q \otimes Q^{\frac{1}{2}}(\Sigma^*-\lambda_i I_k)^+Q^{\frac{1}{2}})+\frac{1}{2}\frac{1}{\lambda_i}\alpha_{pf,i}(\alpha'_{pf,i}Q \otimes \alpha'_{pf,i}Q)$$

Principal Components (PC):

(i) $\min_{A} E[\min_{\upsilon} (X-A\upsilon)'(X-A\upsilon)]$

(ii) $\Sigma A_{pc} = A_{pc}\Lambda_{pc,max,q}$ and $A'_{pc}A_{pc} = I_q$

(iii)   for i=1,...,q:

$$\nabla(\alpha_{pc,i};\Sigma) = -(\alpha'_{pc,i} \otimes (\Sigma-\lambda_{pc,i}I_k)^+)$$

PR coefficient of linear association:

(a) $$\rho^2_{pr} = 1 - \frac{tr \Lambda_{min,p}}{tr \Lambda_k}$$

(b) $$\nabla(\rho^2_{pr};\Sigma) = \frac{(vec((1-\rho^2_{pr})Q - B_{pr}B'_{pr}))'}{tr \Lambda_k}$$

PF coefficient of linear association:

(a)
$$\rho^2_{pf} = \frac{\text{tr } \Lambda_{max,q}}{\text{tr } \Lambda_k}$$

(b)
$$\nabla(\rho^2_{pf}; \Sigma) = \frac{(\text{vec}(QA_{pf}\Lambda^{-1}_{max,q}A'_{pf}Q - \rho^2_{pf}Q))'}{\text{tr } \Lambda_k}$$

SE coefficient of linear association:

(a)
$$\rho^2_{se} = 1 - \frac{\text{tr } \Omega_{min,p}}{\text{tr}(Q_{yy}\Sigma_{yy})}$$

(b)
$$\nabla(\rho^2_{se}; \Sigma) = \frac{(\text{vec}((1-\rho^2_{se}) \begin{bmatrix} Q_{yy} & 0 \\ 0 & 0 \end{bmatrix} - B_{se}B'_{se}))'}{\text{tr}(Q_{yy}\Sigma_{yy})}$$

SUR coefficient of linear association:

(a)
$$\rho^2_{sur} = \frac{\text{tr}(Q_{yy}\Sigma'_{zy}\Sigma^{-1}_{zz}\Sigma_{zy})}{\text{tr}(Q_{yy}\Sigma_{yy})}$$

(b)
$$\nabla(\rho^2_{sur}; \Sigma) = \frac{(\text{vec} \begin{bmatrix} -\rho^2_{sur}Q_{yy} & 0 \\ 2B_{sur,z}Q_{yy} & -B_{sur,z}Q_{yy}B'_{sur,z} \end{bmatrix})'}{\text{tr}(Q_{yy}\Sigma_{yy})}$$

VII. Computational Aspects of the Population-Sample Decomposition

VII.1 Fourth-order central moments

The distinction between population and sample aspects of the estimation problem has its advantage in practical applications. This fact is once more emphasized by the general form of the summary in the preceding review. However, when programming this technique one may encounter some efficiency problems concerning the use of computer space and time, as this method demands the computation of fourth-order moments. In this section the practical aspects of the limiting distribution of $\text{vec}\hat{\Sigma}$, and especially its asymptotic covariance matrix V, will be considered. As the components of μ, Σ and V have to be estimated element by element, it is convenient to recall the notational conventions about the components of X and its various moments. For the definitions of these notations one is referred to section II.2 (see (II.2.11)-(II.2.15)), which are, in short, summarized by

(1.1)
$$X = (X_1,\ldots,X_k)'$$
$$\mu_i = E(X_i)$$
$$\sigma_{ij} = E(X_i-\mu_i)(X_j-\mu_j)$$
$$\upsilon_{ihj\ell} = E(X_i-\mu_i)(X_h-\mu_h)(X_j-\mu_j)(X_\ell-\mu_\ell)$$

for $i,h,j,\ell=1,\ldots,k$. Using this notation, the $((h-1)k+i, (\ell-1)+j)^{th}$ element of V is given by

(1.2)
$$V_{ih,j\ell} = \upsilon_{ihj\ell} - \sigma_{ih}\sigma_{j\ell}.$$

When the sample consists of N i.i.d. observations $\{X_n: n=1,\ldots,N\}$, then the sample analogues of the parameters denoted in (1.1) are given by

(1.3)
$$\hat{\mu}_i = \frac{1}{N} \sum_{n=1}^{N} X_{n,i}$$

$$\hat{\sigma}_{ij} = \frac{1}{N} \sum_{n=1}^{N} (X_{n,i}-\hat{\mu}_i)(X_{n,j}-\hat{\mu}_j)$$

$$\hat{\upsilon}_{ihj\ell} = \frac{1}{N} \sum_{n=1}^{N} (X_{n,i}-\hat{\mu}_i)(X_{n,h}-\hat{\mu}_h)(X_{n,j}-\hat{\mu}_j)(X_{n,\ell}-\hat{\mu}_\ell).$$

A straightforward estimator of $V_{ih,j\ell}$, say $\hat{V}_{ih,j\ell}$, is given by

(1.4) $\qquad \hat{V}_{ih,j\ell} = \hat{\upsilon}_{ihj\ell} - \hat{\sigma}_{ih}\hat{\sigma}_{h\ell}.$

Observing (1.4) we distinguish two complications in estimating the fourth-
order central moments. Firstly, all the observations in the sum that defines
$\hat{\upsilon}_{ihj\ell}$ are used in deviation of their sample mean. This implies that in this
estimation approach the elements of $\hat{V}$ have to be computed by going twice
through the set of observations. In the first run the observations X_n are
summed over n, so that the sample mean $\hat{\mu}$ can be obtained. Then, in the second
run, the fourth-order cross-products of the observations can be summed in
deviation of the sample mean. As the input procedures of records require a
relatively long computer time (compared with internal central processor unit
activities, especially when the data is on tape), it is desirable to derive a
modified estimator of V such that only one run through the data file is
necessary.

A second complication is the huge amount of computer space needed to store all
the k^4 fourth-order central moments. For instance, when the vector X is of
length 10, then $\hat{\upsilon}_{ihj\ell}$ is needed 10.000 times, in order to obtain an estimator
of V. However, note that $\hat{\upsilon}_{ihj\ell}$ represents the same value for all permutations
of the indices, i, h, j and ℓ. For instance $\hat{\upsilon}_{ihj\ell} = \hat{\upsilon}_{h\ell ji}$. Hence when all the
subscripts are different, then $\hat{\upsilon}_{ihj\ell}$ is computed 24 times. In matrix
terminology this can be formulated as follows. Considering $\hat{\upsilon}_{ihj\ell}$ as an element
of a four-dimensional (k×k×k×k) array, all the two-dimensional "slices" form a
symmetric (k×k) matrix. Consequently, there is symmetry in four dimensions and
the computed and stored number of elements can be reduced drastically, when we
make use of this symmetry.

Although the two aforementioned complications are described for the fourth-
order central moments, these problems are also met in connection with the
elements of $\hat{\Sigma}$. That is, also for $\hat{\sigma}_{ij}$ are the observations needed in deviation
of their mean and for Σ we have symmetry "in two dimensions". For these
problems in the context of second-order central moments there are well-known
shortcuts. By writing

(1.5) $\qquad \hat{\sigma}_{ij} = \frac{1}{N} \sum_{n=1}^{N} X_{n,i} X_{n,j} - \hat{\mu}_i \hat{\mu}_j$

it follows that $\hat{\sigma}_{ij}$ can be obtained from the non-central (first- and second-order) moments only. Hence the observations are no longer needed in deviation of their sample mean and σ_{ij} can be estimated in one run. About the symmetry of $\hat{\Sigma}$ the following can be said. As $\hat{\sigma}_{ij} = \hat{\sigma}_{ji}$, it is useless to compute both versions of this sample covariance and it suffices to store the $\frac{1}{2}k(k+1)$ upper- and on-diagonal elements only. For instance in case of k=3, the non-redundant elements of the (3×3) matrix $\hat{\Sigma}$ are stored in a (6×1) column-vector $(\hat{\sigma}_{11}, \hat{\sigma}_{12}, \hat{\sigma}_{22}, \hat{\sigma}_{13}, \hat{\sigma}_{23}, \hat{\sigma}_{33})'$. Denoting the position of $\hat{\sigma}_{ij}$ in the vector of non-redundant elements by "pos(i,j)", we have for k=3: pos(1,1)=1, pos(1,2)=2, ..., pos(3,3)=6. For a general (k×k) matrix $\hat{\Sigma}$ the distinct elements with subscripts $1 \leq i \leq j \leq k$ are renumbered as

(1.6) $pos(i,j) = \frac{1}{2}j(j-1)+i.$

Or, in words, vectorizing the non-redundant elements of $\hat{\Sigma}$, element $\hat{\sigma}_{ij} = \hat{\sigma}_{ji}$ with $i \leq j$ can be found on the $(\frac{1}{2}j(j-1)+i)^{th}$ position of the column-vector.

So far for the second-order sample moments. Let us return to the complications as mentioned originally in connection with $\hat{\upsilon}_{ihj\ell}$. Similar to expression (1.5) for $\hat{\sigma}_{ij}$, also $\hat{\upsilon}_{ihj\ell}$ can be rewritten in non-central moments only. Denoting the various non-central moments by

(1.7) $\hat{\mu}_{ij} = \frac{1}{N} \sum_{n=1}^{N} X_{n,i}X_{n,j}$

$\hat{\mu}_{ihj} = \frac{1}{N} \sum_{n=1}^{N} X_{n,i}X_{n,h}X_{n,j}$

$\hat{\mu}_{ihj\ell} = \frac{1}{N} \sum_{n=1}^{N} X_{n,i}X_{n,h}X_{n,j}X_{n,\ell}$

we may rewrite $\hat{\upsilon}_{ihj\ell}$ as

(1.8) $\hat{\upsilon}_{ihj\ell} = \hat{\mu}_{ihj\ell} - \hat{\mu}_i\hat{\mu}_{hj\ell} - \hat{\mu}_h\hat{\mu}_{ij\ell} - \hat{\mu}_j\hat{\mu}_{ih\ell} - \hat{\mu}_\ell\hat{\mu}_{ihj}$

$+ \hat{\mu}_i\hat{\mu}_h\hat{\mu}_{j\ell} + \hat{\mu}_i\hat{\mu}_j\hat{\mu}_{h\ell} + \hat{\mu}_i\hat{\mu}_\ell\hat{\mu}_{hj} + \hat{\mu}_h\hat{\mu}_j\hat{\mu}_{i\ell} + \hat{\mu}_h\hat{\mu}_\ell\hat{\mu}_{ij} + \hat{\mu}_j\hat{\mu}_\ell\hat{\mu}_{ih}$

$- 3 \hat{\mu}_i\hat{\mu}_h\hat{\mu}_j\hat{\mu}_\ell.$

The moments denoted in (1.7) can be computed in one run through the data, so that also $\hat{\upsilon}_{ihj\ell}$ is obtained in one round.

The symmetry of the $(k \times k \times k \times k)$ fourth-order central moments matrix yields some more difficulties. Not only the non-redundant elements have to be placed in a vector, but also a function has to be defined, that yields the position of $\hat{\upsilon}_{ihj\ell}$ in this vector, resulting from the subscripts i, h, j and ℓ. Let us consider, similar to the two-dimensional case, only the elements for which the subscripts are in non-descending order: $i \leq h \leq j \leq \ell$. These elements are arranged such that the first index runs slow and the last index runs fast. Formally, denoting the position number of the element with subscripts $i \leq h \leq j \leq \ell$ by "pos(i,h,j,ℓ)", we may define the following algorithm (in PASCAL), that assigns the position number of $\hat{\upsilon}_{ihj\ell}$ in the vector of non-redundant elements to pos(i,h,j,ℓ).

(1.9)
```
index:=0;
for ℓ:=1 to k do
begin for j:=1 to ℓ do
        begin for h:=1 to j do
                begin for i:=1 to h do
                        begin index:=index+1;
                              pos(i,h,j,ℓ):=index
                        end;
                end;
        end;
end.
```

For example, when k=3, then the order of the subscripts of the non-redundant fourth-order central moments is given by 1111, 1112, 1122, 1222, 2222, 1113, 1123, 1223, 2223, 1133, 1233, 2233, 1333, 2333, 3333. The procedure described in (1.9) is useful for computation and storage of all different fourth-order moments. But what about retrieving the value of $\hat{\upsilon}_{ihj\ell}$, once the values have been stored on place pos(i,h,j,ℓ) for arbitrary index combination $i \leq h \leq j \leq \ell$? For this problem algorithm (1.9) is not suitable, as it would require going through the sequence of indices that precedes the index combination i,h,j,ℓ in question. This retrieval problem has a combinatorial character and it will be dealt with in the next theorem. Before the theorem is presented, first the factorial and combinatorial symbols used in the exposition of the theorem will be provided.

For a positive integer n, the product of all the positive integers smaller than or equal to it, is denoted by n! (read: "n factorial"). Hence

(1.10) $n! = n(n-1)(n-2)...1$

and by definition $0! = 1$. The combinatorial symbol $\binom{n}{k}$ denotes the number of combinations of k drawings out of a group of n objects, and is defined for $0 \leq k \leq n$ by

(1.11) $\binom{n}{k} = \frac{n!}{k!(n-k)!}.$

When $k < 0$ or $k > n$, then by definition $\binom{n}{k} = 0$. A property of these so-called binomial coefficients, which will be used in the proof of the theorem, is given by

(1.12) $\binom{n}{n} + \binom{n+1}{n} + ... + \binom{n+m}{n} = \binom{n+m+1}{n+1}.$

The proof of this result, which is simply obtained by application of mathematical induction with respect to m, is left to the reader. In the following theorem, the problem of ordering and retrieving the elements of a four-dimensional array with symmetry in all four dimensions is reformulated for a more general m-dimensional array in combinatorial terminology.

Theorem (m-dimensional symmetric matrices)
Let $M_{m;k}$ be an m-dimensional (k×k...×k) array for which $M_{m;k}(i_1,...,i_m)$, which denotes the $(i_1,...,i_m)^{th}$ entry of $M_{m;k}$, is the same for all permutations of subscripts $i_1,...,i_m$. Denoting the number of different elements in $M_{m;k}$ by $Num(M_{m;k})$ then

(1.13) $Num(M_{m;k}) = \binom{m+k-1}{m}$

and an enumeration of the different elements is given for $1 \leq i_1 \leq ... \leq i_m \leq k$ by

$$(1.14) \quad \text{pos}_m(i_1,\ldots,i_m) = \binom{i_m+m-2}{m} + \binom{i_{m-1}+m-3}{m-1} + \ldots + \binom{i_1-1}{1} + 1$$

where $\text{pos}_m(i_1,\ldots,i_m)$ takes the values $1,2,\ldots,\text{Num}(M_{m;k})$.

Proof

The different elements of $M_{m;k}$ are given by $M_{m;k}(i_1,\ldots,i_m)$ for which $1 \leq i_1 \leq \ldots \leq i_m \leq k$. Hence the set of subscripts in which we are interested is given by

$$(1.15) \quad I = \{(i_1,\ldots,i_m) \mid 1 \leq i_1 \leq \ldots \leq i_m \leq k\}.$$

Let I be the domain of a function ϕ which is defined by

$$(1.16) \quad \phi(i_1,\ldots,i_m) = (i_1, i_2+1,\ldots,i_m+m-1),$$

then the range of ϕ is given by

$$(1.17) \quad I' = \{(j_1,\ldots,j_m) \mid 1 \leq j_1 < \ldots < j_m \leq m+k-1\}.$$

As the function ϕ is "on" (that is $\phi(I) = I'$) and "1-1" (which means that $(i_1,\ldots,i_m) \neq (h_1,\ldots,h_m)$ both in I, implies that $\phi(i_1,\ldots,i_m) \neq \phi(h_1,\ldots,h_m)$) it follows that the number of elements in I equals the number of elements in I'. Hence the number of elements in I' equals $\text{Num}(M_{m;k})$. As the combination $(j_1,\ldots,j_m)$ in I' can be obtained by selecting m numbers from the set $\{1,2,\ldots,m+k-1\}$ it follows from the definition of the binomial coefficient (see (1.11)) that I' consists of $\binom{m+k-1}{m}$ different elements, which concludes the proof of (1.13).

In order to prove (1.14), it is assumed that the elements with indices $1 \leq i \leq \ldots \leq i_m \leq k$ are ordered according to the procedure as presented for k=4 in (1.9). This means that for $\ell = m, m-1,\ldots,1$ subscribt i_ℓ is not raised until all the preceding subscripts $i_1 \ldots, i_{\ell-1}$ are equal to the value of i_ℓ. The position of $M_{m;k}(i_1 \ldots,i_m)$ as denoted by (1.14) is derived by applying mathematical induction with respect to m (the number of dimensions). We start by observing that the proposed position number holds for m=1:

$$(1.18) \quad \text{pos}_1(i_1) = \binom{i_1-1}{1} + 1 = i_1.$$

The correctness of this statement is trivial as $M_{1;k}$ itself is the vector without any double elements. Now we have to show that (1.14) is true, given that the proposition holds for an $(m-1)$-dimensional matrix $M_{m-1;k}$. The value of $pos_m(i_1,\ldots,i_m)$ will be obtained by counting the number of its predecessors. Let the indices of the elements that preceed $M_{m;k}(i_1,\ldots,i_m)$ be denoted by $j_1,\ldots,j_m$. For the $j_1,\ldots,j_m$ combinations of which j_m is equal to ℓ, where $1 \leq \ell \leq i_m-1$, it is known that they can take all values, such that $j_1 \leq \cdots \leq j_m \leq \ell$. According to (1.13), the number of different elements with $j_m=\ell$ is given by $Num(M_{m-1;\ell})$. Summation over ℓ yields

$$(1.19) \qquad \sum_{\ell=1}^{i_m-1} Num(M_{m-1;\ell}) = \binom{m-1}{m-1} + \binom{m}{m-1} + \ldots + \binom{m+i_m-3}{m-1} = \binom{m+i_m-2}{m}.$$

The last equality is obtained by applying (1.12). Furthermore we have to count the number of elements for which the mth index equals i_m. According to the induction hypothesis, the number of elements with index $(1,\ldots,1,i_m)$ up to index $(i_1,\ldots,i_{m-1},i_m)$ is given by

$$(1.20) \qquad pos_{m-1}(i_1,\ldots,i_{m-1}) = \binom{i_{m-1}+m-3}{m-1} + \binom{i_{m-2}+m-4}{m-2} + \ldots + \binom{i_1-1}{1} + 1.$$

Addition of the number of elements counted in (1.19) and in (1.20) indeed yields the position number as proposed in (1.14). $\diamond$

What does the preceding theorem mean for the second- and fourth-order central moments? Reformulation of (1.13) and (1.14) for $m=2$ yields the formulas for $\hat{\sigma}_{ij}$:

$$(1.21) \qquad Num(M_{2;k}) = \binom{k+1}{2} = \tfrac{1}{2}k(k+1)$$

$$(1.22) \qquad pos_2(i,j) = \tfrac{1}{2}j(j-1) + i,$$

which corresponds with the results as denoted in (1.6). Reformulation of (1.13) and (1.14) for $m=4$ yields the formulas for $\hat{\upsilon}_{ihj\ell}$:

(1.23) $\quad \text{Num}(M_{4;k}) = \binom{k+3}{4} = \frac{1}{24} k(k+1)(k+2)(k+3)$

(1.24) $\quad \text{pos}_4(i,h,j,\ell) = i+\frac{1}{2}[(h-1)h + \frac{1}{3}[(j-1)j(j+1)+\frac{1}{4}(\ell-1)\ell(\ell+1)(\ell+2)]].$

Consequently, only non-redundant values of $\hat{\upsilon}_{ihj\ell}$ have to be stored, so that the computer space, which is necessary for the fourth-order central moments, is reduced from k^4 to $\frac{1}{24}k(k+1)(k+2)(k+3)$ and given any arbitrary index sequence $i \leq h \leq j \leq \ell$, the corresponding element can be retrieved by means of the function as proposed in (1.24).

VII.2 Pre- and post-multiplication of V by the gradient matrix

The efficient storage structure for the fourth-order central moments $\hat{\upsilon}_{ihj\ell}$ as described in the preceding section, yields a substantial reduction of the computer space that is necessary to store the information. However, what are the consequences of this storage method for the practical use of the PSD-approach? Recall that $\upsilon_{ihj\ell}$ forms an element of the middle matrix V in $\nabla V \nabla'$, where ∇ represents the matrix of partial derivatives of the parameter vector of interest, with respect to $(\text{vec}\Sigma)'$. (note that ∇ is merely a short-hand notation for the matrices $\nabla(B_{.};\Sigma)$, $\nabla(A_{.};\Sigma)$ and $\nabla(\rho_{.}^2;\Sigma)$, introduced in chapters III, IV and V). We are interested in an estimator of this matrix product which can be obtained by replacing V by $\hat{V}$ and ∇ by $\hat{\nabla}$, where $\hat{\nabla}$ is defined as the method of moments estimator for ∇. Denoting the $(r, (h-1)k+i)^{th}$ element of ∇ (which represents the derivative of the rth component of the parameter vector with respect to σ_{ih}) by $\nabla_{r,ih}$, it follows that an estimator of the $(r,s)^{th}$ element of $\nabla V \nabla'$ is given by

(2.1) $\qquad [\hat{\nabla}\hat{V}\hat{\nabla}']_{rs} = \sum\limits_{i=1}^{k} \sum\limits_{h=1}^{k} \sum\limits_{j=1}^{k} \sum\limits_{\ell=1}^{k} \hat{\nabla}_{r,ih}(\hat{\upsilon}_{ihj\ell}-\hat{\sigma}_{ih}\hat{\sigma}_{j\ell})\hat{\nabla}_{s,j\ell}.$

Hence, $\hat{\upsilon}_{ihj\ell}$ is needed for all k^4 possible values of the sequence i, h, j, ℓ in the computation of one element of the asymptotic covariance matrix. Naturally the formula as denoted in (1.24) could be used in order to locate each $\hat{\upsilon}_{ihj\ell}$. This would yield two activities:

(i)   arrangement of subscripts i, h, j, ℓ in a non-descending sequence

(ii)   computation of the value of $\text{pos}_4(i,h,j,\ell)$ for $i \leq h \leq j \leq \ell$.

For a single element $\hat{\upsilon}_{ihj\ell}$ this procedure can easily be performed. However, as one may conclude from (2.1), these actions have to be performed for all k^4 combinations of subscripts, which might be very computer intensive, especially when k is large. Let us therefore leave this retrieval method and concentrate ourselves on the procedure introduced in (1.9).

Going through the sequence of subscripts as denoted in (1.9), we only encounter those values for which $i \leq h \leq j \leq \ell$. Therefore this algorithm is in its present form not yet applicable, as it does not give any information about how many times each element $\hat{\upsilon}_{ihj\ell}$ has to be used in summation (2.1). For instance when the subscripts i, h, j, ℓ all have different values then the corresponding element $\hat{\upsilon}_{ihj\ell}$ appears 24 times in the summation, while the value for $\hat{\upsilon}_{ihj\ell}$ with $i = h = j = \ell$ appears only once. Therefore we shall, before extending the algorithm to apply it for computation of the sum denoted in (2.1), analyse the number of different subscript permutations when one or more of the $\leq$-signs in fact denote an inequality so that "$\leq$" can be replaced by "$<$". The various possible combinations of "$<$" and "$=$" in combination with the subscripts are given in the first column of table VII.2.1.

i,h,j,ℓ combinations	binary code	decimal code
$i < h < j < \ell$	000	0
$<$ $<$ $=$	001	1
$<$ $=$ $<$	010	2
$<$ $=$ $=$	011	3
$=$ $<$ $<$	100	4
$=$ $<$ $=$	101	5
$=$ $=$ $<$	110	6
$=$ $=$ $=$	111	7

Table VII.2.1 Codification of subscript sequences

Neglecting the subscripts i,h,j,ℓ and replacing $<$ by 0 and $=$ by 1, then a binary code is obtained of three 0-1 digits. These 8 binary codes, that determine the $<$, $=$ combinations uniquely, are given in the second column of table VII.2.1 and they can be considered as the binary representation of the numbers $0,1,\ldots,7$ which are given in the third column in their decimal representation. Consequently, the permutations of the subscripts that yield different subscript sequences can be determined by the value of the decimal code (which shall from now on in short be referred to by "code"). Combining the procedure given in (1.9) with the binary code representation technique, we can obtain an algorithm that goes through all the values of sequence i, h, j, ℓ for which $i \leq h \leq j \leq \ell$ and which gives the corresponding index-values together with their code-values

```
(2.2)    index:=0;
         code:=0;
         for ℓ:=1 to k do
         begin for j:=1 to ℓ do
                begin if ℓ=j then code:=code+1;
                       for h:=1 to j do
                       begin if h=ℓ then code:=code+2;
                              for i:=1 to h do
                              begin if i=h then code:=code+4;
                                     index:=index+1;
                                     .

                                     .

                                     {computations according to (2.1)}

                                     .

                                     .

                              end;
                              code:=code-4
                       end;
                       code:=code-2
                end;
                code:=code-1
         end.
```

On the place in the algorithm indicated by the comment "{computations according to (2.1)}", the values of i, h, j, ℓ, index and code are used for the actual computation of the pre- and post-multiplication. Before we come to filling up this program part, first the different index sequences, given the value of code, will be inspected. Equality of the subscripts i and h, say, implies that one should not use the combination of subscripts in which only the order of i and h are reversed.

Let us consider as an example the case that code=6. Then, according to tabel VII.2.1, we have $i=h=\ell$. Consequently, the permutation of these three subscripts only, does not yield any new combinations. Hence we are restricted to the permutations with i before h and h before j, yielding the 4 combinations (i,h,j,ℓ), (i,h,ℓ,j), (i,ℓ,h,j) and (ℓ,i,h,j). In table VII.2.2 all possible permutations for the various values of "code" are given.

code	possible permutations of (i,h,j,ℓ)
7	(i,h,j,ℓ)
6	(i,h,ℓ,j), (i,ℓ,h,j), (ℓ,i,h,j), and permutation for code=7
5	(i,j,h,ℓ), (j,i,h,ℓ), (j,i,ℓ,h), (j,ℓ,i,h), (i,j,ℓ,h), and permutation for code=7
4	(ℓ,j,i,h), (i,ℓ,j,h), (ℓ,i,j,h), and permutations for code=7, code=6 and code=5
3	(h,i,j,ℓ), (h,j,i,ℓ), (h,j,ℓ,i), and permutation for code=7
2	(ℓ,h,j,i), (h,ℓ,j,i), (ℓ,h,i,j), (h,i,ℓ,j), (h,ℓ,i,j), and permutations for code=7, code=6 and code=3
1	(j,ℓ,h,i), (j,h,ℓ,i), (j,h,i,ℓ), and permutations for code=7, code=5 and code=3
0	(ℓ,j,h,i), and all non-redundant permutations for code=1,...,7

Table VII.2.2 Subscript sequences for the various values of the code

Note that table VII.2.2 indeed forms a list of all 24 permutations of the subscripts, which occur in case code=0.

Using the information of table VII.2.2, procedure (2.2) can be completed in such a way that an algorithm is obtained that computes (2.1). A function is defined that computes, given the values of r, s, i', h', j' and ℓ', one term $\hat{\nabla}_{r,i'h'}(\hat{\upsilon}_{i'h'j'\ell'}-\hat{\sigma}_{i'h'}\hat{\sigma}_{j'\ell'})\hat{\nabla}_{s,j'\ell'}$, where the indices i', h', j', ℓ' represent a permutation of the subscripts i, h, j, ℓ, which are not necessarily in a non-descending order. The function will be given the name PRODUCT, indicating the fact that an element of $\hat{V}$ is pre- and post-multiplied by elements of $\hat{\nabla}$. The vectors containing the non-redundant second-order and fourth-order central moments $\hat{\sigma}_{ij}$ and $\hat{\upsilon}_{ihj\ell}$ are in the algorithm denoted by the arrays SIGMA and NU respectively. The gradient matrix $\hat{\nabla}$ is denoted by the 2-dimensional array DELTA. In order to retrieve the elements $\hat{\sigma}_{i'h'}$ and $\hat{\sigma}_{j'\ell'}$ in array SIGMA, the method given in (1.22) is used. That is, the place of $\sigma_{i'h'}$ in SIGMA is given by $\frac{1}{2}i'(i'-1)+h'$ if $h' \leq i'$, or by $\frac{1}{2}h'(h'-1)+i'$ if $h' > i'$. The values of the relevant indices are in the function stored in the local integer variables "ind1" and "ind2". The declaration function PRODUCT is given by

(2.3)   <u>function</u> PRODUCT(i',h',j',ℓ': <u>integer</u>): <u>real</u>;

      <u>begin</u> ind1, ind2: <u>integer</u>;

          <u>if</u> h'$\leq$i' <u>then</u> ind1:=i'*(i'-1)/2+h' <u>else</u> ind1:=h'*(h'-1)/2+i';

          <u>if</u> j'$\leq$ℓ' <u>then</u> ind2:=ℓ'*(ℓ'-1)/2+j' <u>else</u> ind2:=j'*(j'-1)/2+ℓ';

          . PRODUCT := DELTA[r,(i'-1)*k+h']*(NU[index]-SIGMA[ind1]*SIGMA[ind2])

                *DELTA[s,(ℓ'-1)*k+j']

      <u>end</u>;

Combining the subscript permutations for a given value of "code" (given in table VII.2.1) with the function PRODUCT, we are able to replace the comment in algorithm (2.2) by the relevant programming commands, so that a complete computer procedure is obtained that computes the $(r,s)^{th}$ element of $\hat{\nabla}\hat{\nabla}\hat{\nabla}'$. Denoting this element by COVAR[r,s], as it represents the $(r,s)^{th}$ element of the asymptotic covariance matrix, this part of the algorithm is given by

(2.4)   COVAR[r,s] := PRODUCT(i,h,j,ℓ);

      <u>if</u> code <= [0,2,4,6] <u>then</u> COVAR[r,s] := COVAR[r,s] +

          PRODUCT(i,h,ℓ,j) + PRODUCT(i,ℓ,h,j) + PRODUCT(ℓ,i,h,j);

      <u>if</u> code <= [0,1,4,5] <u>then</u> COVAR[r,s] := COVAR[r,s] +

          PRODUCT(i,j,h,ℓ) + PRODUCT(j,i,h,ℓ) + PRODUCT(j,i,ℓ,h) +

          PRODUCT(j,ℓ,i,h) + PRODUCT(i,j,ℓ,h);

      <u>if</u> code <= [0,4] <u>then</u> COVAR[r,s] := COVAR[r,s] +

          PRODUCT(ℓ,j,i,h) + PRODUCT(i,ℓ,j,h) + PRODUCT(ℓ,i,j,h);

      <u>if</u> code <= [0,1,2,3] <u>then</u> COVAR[r,s] := COVAR[r,s] +

          PRODUCT(h,i,j,ℓ) + PRODUCT(h,j,i,ℓ) + PRODUCT(h,j,ℓ,i);

      <u>if</u> code <= [0,2] <u>then</u> COVAR[r,s] := COVAR[r,s] +

          PRODUCT(ℓ,h,j,i) + PRODUCT(h,ℓ,j,i) + PRODUCT(ℓ,h,i,j) +

          PRODUCT(h,i,ℓ,j) + PRODUCT(h,ℓ,i,j);

      <u>if</u> code <= [0,1] <u>then</u> COVAR[r,s] := COVAR[r,s] +

          PRODUCT(j,ℓ,h,i) + PRODUCT(j,h,ℓ,i) + PRODUCT(j,h,i,ℓ);

      <u>if</u> code <= [0] <u>then</u> COVAR[r,s] := COVAR[r,s] + PRODUCT(ℓ,j,h,i);

(The expression "code <= [0,2,4,6]" is the PASCAL notation for "code ϵ {0,2,4,6}"). Performing the algorithm, that is obtained after combination of (2.2), (2.3) and (2.4), for all possible values of (r,s), the complete asymptotic covariance matrix of the PSD estimator is computed. Naturally this covariance matrix is symmetric, so that also here the efficient storage method

as explained in section 1 can be applied. For instance, in the PR case
$\frac{1}{2}pk(pk+1)$ and in the PF case $\frac{1}{2}qk(qk+1)$ elements have to be computed, where pk
and qk are the numbers of elements in matrix B_{pr} and A_{pf} respectively.

For estimation of $B_.$, $A_.$, $\rho_.^2$ and the corresponding gradient matrices, the
application of more elementary manipulations of symmetric two-dimensional
matrices, like computation of matrix products, matrix inverses, characteristic
roots and characteristic vectors are needed. The efficient storage method for
symmetric matrices yields adaptation of computer algorithms for these
operations, which can for instance be found in the IBM application program
SYSTEM/360 Scientific Subroutine Package (SSP). The IBM Programmers Manual
(1970) contains listings and sufficient information to permit the reader to
understand the working of this kind of procedures.

VII.3 The PSD method in practice.

In order to observe the performance of the PSD method in practice, some of the minimum distance parameters that have been analysed in the preceding chapters are computed, using a large dataset that forms a cross-section of Dutch families in 1977. In this section we shall consider two samples that are obtained as subsets of the aforementioned dataset.

The first sample consist of 2206 independent and identically distributed (i.i.d.) observations of a vector with three variables (that is $N = 2206$ and $k = 3$), denoted by

$$(3.1) \qquad X = \begin{bmatrix} \ln(y_{min}) \\ \ln(y_c) \\ \ln(fs) \end{bmatrix},$$

where "y_{min}" stands for the absolute minimum income level (in Dutch guilders) a family would need to run their household, "y_c" for the current after-tax income and "fs" for the family size. This sample has already been used in the context of functional linear regression analysis. The results of OLS regression analysis for this example can for instance be found in Van Praag, Goedhart, Kapteyn (1980). For the theoretical backgrond of the relation in which $\ln(y_c)$ and $\ln(fs)$ are treated as independent variables and $\ln(y_{min})$ as the variable to be explained, one is also referred to Van Praag (1971) and Goedhart, Halberstadt, Kapteyn, Van Praag (1977).

Independent of the parameters that have to be estimated, we may compute the different-order sample moment, yielding for instance the consistent estimators of the mean vector μ, covariance matrix Σ and the matrix of fourth-order central moments, say Π. The results for our example are given by

$$(3.2) \qquad \hat{\mu} = \begin{bmatrix} 9.842 \\ 10.072 \\ 1.024 \end{bmatrix}, \quad \hat{\Sigma} = \begin{bmatrix} 0.147 & 0.124 & 0.093 \\ & 0.241 & 0.097 \\ & & 0.279 \end{bmatrix},$$

$$\hat{\Pi}_{1111} = 0.101 \qquad \hat{\Pi}_{1113} = 0.052 \qquad \hat{\Pi}_{1233} = 0.060$$

$$\hat{\Pi}_{1112} = 0.075 \qquad \hat{\Pi}_{1123} = 0.046 \qquad \hat{\Pi}_{2233} = 0.085$$

$$\hat{\Pi}_{1122} = 0.083 \qquad \hat{\Pi}_{1223} = 0.050 \qquad \hat{\Pi}_{1333} = 0.070$$

$$\hat{\Pi}_{1222} = 0.083 \qquad \hat{\Pi}_{2223} = 0.042 \qquad \hat{\Pi}_{2333} = 0.076$$

$$\hat{\Pi}_{2222} = 0.354 \qquad \hat{\Pi}_{1133} = 0.064 \qquad \hat{\Pi}_{3333} = 0.197$$

Combination of $\hat{\Sigma}$ and $\hat{\Pi}$ yields an estimator of V (the covariance matrix of $\mathrm{vec}(XX')$) as

(3.3) $V = \Pi - (\mathrm{vec}\Sigma)(\mathrm{vec}\Sigma)'$,

yielding the solution of the sample problem in case of an ideal sample consisting of i.i.d. observations. The moments μ, Σ and Π are parameters of the distribution of X in three dimensions. Applying PR analysis with p=1 or PF analysis with q=2, one can estimate the parameters of the two-dimensional linear subspace that fits the set of observations as good as possible. This linear subspace is described by the relation

(3.4) $\beta'_{pr}\hat{X} = \gamma$ or $\hat{X} = A_{pf}U$

with β_{pr} a (3×1) vector, γ a scalar and A_{pf} a (3×2) matrix. The parameter γ is introduced, as in this example the expectation of X does not equal zero, like it was assumed in the theory. A non-zero expectation yields a translation of the optimal linear subspace, such that it passes through the central point of the observation set. An estimator of γ can be obtained by observing that

(3.5) $\beta'_{pr}(\hat{X} - \mu) = 0$

so that γ is consistently estimated by

(3.6) $\hat{\gamma}_{pr} = \hat{\beta}'_{pr}\hat{\mu}.$

Observing the minimum distance problem in the metric defined by $Q = I_3$, the values of the parameters for the PR case are given by

$$(3.7) \qquad \hat{\beta}_{pr} = \begin{bmatrix} 0.819 & (0.016) \\ -0.567 & (0.027) \\ -0.094 & (0.024) \end{bmatrix} \qquad \text{and} \quad \hat{\gamma}_{pr} = 2.256 \quad (0.415).$$

The numbers within parentheses behind the estimates are the asymptotic standard deviations, that are obtained by taking the square root of the diagonal elements of the asymptotic covariance matrix of $\hat{\beta}_{pr}$. In the PF case the estimator of A_{pf} equals

$$(3.8) \qquad \hat{A}_{pf} = \begin{bmatrix} 0.310 & (0.007) & \vdots & -0.130 & (0.015) \\ 0.376 & (0.011) & \vdots & -0.237 & (0.008) \\ 0.436 & (0.011) & \vdots & 0.297 & (0.006) \end{bmatrix}$$

and the goodness-of-fit measure, both for PR analysis and for PF analysis, is given by

$$(3.9) \qquad \hat{\rho}^2_{pr} = \hat{\rho}^2_{pf} = 0.920 \quad (0.006).$$

The vector $\hat{\beta}_{pr}$ and the columns of $\hat{A}_{pf}$ are normalized in the metric defined by I_3, as proposed in the preceding theory. Note that the value of the goodness-of-fit measure is high. This is an indication that it is unreasonable to compare this measure with the multiple correlation coefficiënt, obtained after applying OLS regression techniques to this set of observations. The possibility of a high value for the goodness-of-fit measure (or in fact, a relatively low expected squared distance between X and $\tilde{X}$) was already noted by Pearson in 1901 (p.571). We shall return to this point after the presentation of the result of the PSD estimators in the regression case.

Applying the PR and PF analysis in the metric defined by I_3 with $p=2$ and $q=1$, which means that the observations are projected onto a one-dimensional linear subspace, then the results are given by

$$(3.10) \quad \hat{B}_{pr} = \begin{bmatrix} 0.819 & (0.016) & \vdots & -0.324 & (0.037) \\ -0.567 & (0.027) & \vdots & -0.324 & (0.021) \\ -0.094 & (0.024) & \vdots & 0.740 & (0.015) \end{bmatrix} , \quad \hat{\gamma}_{pr} = \begin{bmatrix} 2.256 & (0.415) \\ -8.370 & (0.307) \end{bmatrix}$$

$$(3.11) \quad \hat{\alpha}_{pf} = \begin{bmatrix} 0.310 & (0.007) \\ 0.376 & (0.011) \\ 0.436 & (0.011) \end{bmatrix}$$

$$(3.12) \quad \hat{\rho}_{pr}^2 = \hat{\rho}_{pf}^2 = 0.669 \quad (0.012).$$

Naturally the first column of $\hat{B}_{pr}$ equals $\hat{\beta}_{pr}$, as the best-fitting one-dimensional subspace is obtained by taking the intersection of the best-fitting two-dimensional subspace, defined by $\hat{\beta}_{pr}$, and the "second-best-fitting" two-dimensional subspace, defined by the second column of $\hat{B}_{pr}$ (the best-fitting line lies in the best-fitting plane).

Let the vector X be partitioned into an endogenous part Y and an exogenous part Z, such that

$$(3.13) \quad Y = \ln(y_{min}) \quad \text{and} \quad Z = \begin{bmatrix} \ln(y_c) \\ \ln(fs) \end{bmatrix}.$$

Now we are able to perform SUR analysis with p=1. (Although the characters SU are meaningless in this example, as we are dealing with one relation only, the acronym SUR will still be used to indicate the regression results.) That is, the observations are projected parallel to the Y-axis onto the best fitting two-dimensional linear subspace and the first coefficient of the parameter vector corresponding to Y equals -1, yielding the relation

$$(3.14) \quad \tilde{Y} = \gamma_{sur} + \beta'_{sur,z} Z.$$

The estimation results in our example are given by

$$(3.15) \qquad \hat{\beta}_{sur,z} = \begin{bmatrix} 0.508 & (0.034) \\ 0.157 & (0.017) \end{bmatrix}, \qquad \hat{\gamma}_{sur} = 4.563 \quad (0.330)$$

and

$$(3.16) \qquad \hat{\rho}^2_{sur} = 0.527 \quad (0.028).$$

Re-normalizing the values of $\hat{\beta}_{pr}$ and $\hat{\gamma}_{pr}$ in (3.7), such that the first component of $\hat{\beta}_{pr}$ equals -1, changing the sign of $\hat{\gamma}_{pr}$ and denoting the results by $\hat{\beta}^*_{pr}$ and $\hat{\gamma}^*_{pr}$, we obtain

$$(3.17) \qquad \hat{\beta}^*_{pr} = \begin{bmatrix} -1 \\ 0.692 \\ 0.115 \end{bmatrix}, \qquad \hat{\gamma}^*_{pr} = 2.755$$

(the standard deviations are left out of consideration). Comparing (3.17) with $\hat{\beta}_{sur}$ and $\hat{\gamma}_{sur}$ in (3.15) (recall that $\beta_{sur,y} = -1$ by definition), it becomes clear how the position of the best-fitting plane is influenced by a change in the projection direction. The results of the SUR analysis are also easily compared with the OLS results obtained from the linear regression model in which it is assumed that the observations Z_n ($n=1,\ldots,N$) are non-random explanatory variables, such that

$$(3.18) \qquad Y_n = c + b'Z_n + \varepsilon_n \qquad \text{for } n=1,\ldots,N$$

where ε_n is a random disturbance term with

$$(3.19) \qquad E(\varepsilon_n) = 0, \quad E(\varepsilon_n^2) = \sigma_\varepsilon^2, \quad E(\varepsilon_m\varepsilon_n) = 0 \quad \text{for } m \neq n.$$

The OLS results for the parameters of (3.18) are given by

$$(3.20) \quad \hat{b} = \begin{bmatrix} 0.508 & (0.012) \\ 0.157 & (0.013) \end{bmatrix}, \quad \hat{c} = 4.563 \ (0.129).$$

Comparing (3.15) with (3.20) it follows that the only difference is in the (asymptotic) standard deviations of the estimates. The higher values of the standard deviations obtained in the PSD case, may be explained by the fact that the linear relation as hypothesized in (3.18) does not need to be true in reality (see also Van Praag (1981)).

Let us now return to the remarkable high value of $\hat{\rho}^2_{pr}$, which equals 0.920 when $p=1$, while $\hat{\rho}^2_{sur} = 0.527$. In order to get an idea of this large difference between $\hat{\rho}^2_{pr}$ and $\hat{\rho}^2_{sur}$, the definitions of these two parameters are repeated, together with a graphic represention for the two-dimensional situation in the metric defined by I_2 with $X = (Y,Z)'$. As usual the vectors are observed in deviation of the mean, which is in the graph presented by defining the mean of X as the origin.

$$(3.21) \quad \rho^2_{pr} = 1 - \left(\frac{E\|X - \tilde{X}\|_Q}{E\|X\|_Q} \right)^2$$

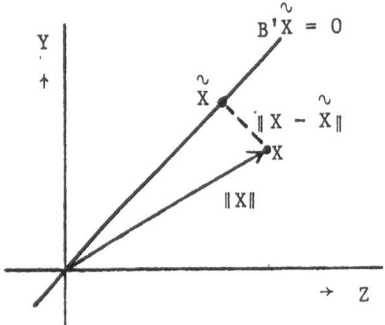

Figure VII.3.1 Graphic representation for the defenition of ρ^2_{pr}.

(3.22) $\rho_{sur}^2 = 1 - \left(\frac{E\|Y - \hat{Y}\|}{E\|Y\|}\right)^2$

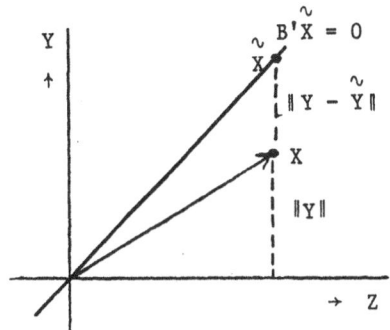

Figure VII.3.2 Graphic representation for the definition of ρ_{sur}^2.

Note the differences in the definitions of ρ_{pr}^2 and ρ_{sur}^2 ; ρ_{sur}^2 is obtained in the same way as ρ_{pr}^2, but after projection of X and $\tilde{X}$ on the subspace spanned by the Y-axis. Why neglecting the Z-dimensions in the definition of ρ_{sur}^2? In the PR case, the goodness-of-fit measure ρ_{pr}^2 considers the norm of $(X-\tilde{X})$ relative to the norm of X. When Q is a two-dimensional diagonal matrix and its second diagonal element tends to infinity, then the projection direction changes and the line segment $(X-\tilde{X})$ turns to the right until it is parallel to the Y-axis. The vector X (and therefore also its norm $\|X\|_Q$) remains unchanged. Consequently, because of the infinite component of Q, ρ_{pr}^2 tends to 1, so that this value is useless in order to measure goodness-of-fit in case of SUR analysis. It was for this reason that for a goodness-of-fit measure for the SUR analysis the norm of $(X-\tilde{X})$, which equals the norm of $(Y-\tilde{Y})$ as $\tilde{Z} = Z$, is considered relative to the norm of Y, instead of the norm of X (see figure VII.3.2). Hence ρ_{sur}^2 does not follow directly from ρ_{pr}^2 when one or more diagonal elements (or in general one or more characteristic roots) of Q tend to infinity. It is this discontinuity in the definitions of the goodness-of-fit measures that causes the large difference between $\hat{\rho}_{pr}^2$ and $\hat{\rho}_{sur}^2$.

As a second example we consider a dataset consisting of 418 observations with 5 variables (hence N = 418 and k = 5). This sample represents the Dutch two-earner families in 1977, hence both the breadwinner and his (or her) partner participate in the labour market. The number of hours worked per week by the head of the household is denoted by H_h and the labour time of the partner is denoted by H_p. Furthermore we have information on the wage rate of the partner (w_p), the family size (fs) and the age of the partner (age_p). The vector of interest is given by

$$(3.23) \quad X = \begin{bmatrix} \ln(H_h) \\ \ln(H_p) \\ \ln(w_p) \\ \ln(fs) \\ \ln(age_p) \end{bmatrix} .$$

More information and economic theory about linear relations between these variables can be found in Homan, Hagenaars, Van Praag (1987), who use these variables (among others) to estimate a time allocation model for household. In the following the estimation results will be given for PR (p=2), PF (q=2) and SUR (p=2) analysis. The discussion of the results will be limited to a minimum, as the methods have been analysed extensively in the theoretical chapters and some remarkable empirical experiences have been dealt with in the preceding example.

Two principal relations describe a three-dimensional linear subspace with minimum expected squared distance to the set of observations, yielding a (5x3) parameter matrix B_{pr} and a (2x1) parameter vector γ_{pr}. Estimating these parameters in the metric defined by I_5, we obtain the following estimates, with the PSD-standard deviations given in parentheses

$$(3.24) \quad \hat{B}_{pr} = \begin{bmatrix} 0.975 & (0.010) & \vdots & 0.037 & (0.138) \\ -0.050 & (0.023) & \vdots & -0.016 & (0.029) \\ 0.106 & (0.033) & \vdots & 0.061 & (0.038) \\ -0.140 & (0.080) & \vdots & -0.492 & (0.046) \\ -0.129 & (0.124) & \vdots & 0.867 & (0.030) \end{bmatrix} , \quad \hat{\gamma}_{pr} = \begin{bmatrix} 3.143 & (0.428) \\ 2.715 & (0.490) \end{bmatrix}$$

and the goodness-of-fit is estimated by

$$(3.25) \quad \hat{\rho}^2_{pr} = 0.909 \quad (0.007).$$

One may think of the vector X as consisting of an endogenous part Y and an exogenous part Z, given by

$$(3.26) \quad Y = \begin{bmatrix} \ln(H_h) \\ \ln(H_p) \end{bmatrix}, \quad Z = \begin{bmatrix} \ln(w_p) \\ \ln(fs) \\ \ln(age_p) \end{bmatrix}.$$

Partitioning the parameter matrix B accordingly and setting the parameter sub-matrix, related with the endogenous variables, equal to $-I_2$, we obtain the SUR case. The estimators of the SUR parameters are for this example given by

$$(3.27) \quad \hat{B}_{sur,z} = \begin{bmatrix} -0.112 & (0.028) & -0.332 & (0.068) \\ 0.090 & (0.034) & -0.638 & (0.095) \\ 0.075 & (0.041) & -0.246 & (0.107) \end{bmatrix}, \quad \hat{\gamma}_{sur} \begin{bmatrix} 3.647 & (0.132) \\ 5.220 & (0.357) \end{bmatrix}$$

and the goodness-of-fit is measured by

$$(3.28) \quad \hat{\rho}^2_{sur} = 0.227 \quad (0.034).$$

This section ends with the results for the PF parameters with q=2 and $Q = I_5$. Hence, the observations are projected onto a two-dimensional subspace and the parameters are estimated by

$$(3.29) \quad \hat{A}_{pf} = \begin{bmatrix} 0.019 & (0.016) & -0.083 & (0.018) \\ 0.702 & (0.020) & 0.104 & (0.040) \\ -0.234 & (0.054) & 0.481 & (0.018) \\ -0.210 & (0.024) & -0.147 & (0.038) \\ -0.091 & (0.020) & -0.112 & (0.021) \end{bmatrix}.$$

The corresponding goodness-of-fit measure equals

$$(3.30) \quad \hat{\rho}^2_{pf} = 0.782 \quad (0.012),$$

which is less than $\hat{\rho}^2_{pf}$ in (3.25), as PR for p=2 yields a fit of the observations in three dimensions, while PF for q=2 fits the observations in two dimensions.

Appendix. Preliminaries on matrix algebra

The formulas in this study are derived and formulated in matrix notation. The applied matrix operators can not always be reckoned among the mathematical techniques that belong to everyone's common knowledge. Therefore the matrix operation properties of which the author thinks that they are important for readability of the chapters II-V, are shortly reviewed in this appendix. It is assumed that the reader is familiar with the "common" matrix manipulations and characteristics, like symmetry, traces, determinants, inverses, characteristic roots and vectors. For a complete treatment of these properties one is referred to matrix analysis books, such as Gantmacher (1959), Bellman (1970) and Golub and Van Loan (1983). The purpose of this appendix is to bring together the more recently developed matrix operations, such as the Kronecker product, the vec-operator, generalized inverses and matrix differentiation techniques. The author is aware of the subjectivity of this choice, but in order to keep this review within limits it seems reasonable to outline only the newer matrix theory which is employed in this study. As usual, all the formulas given in this appendix will be preceded by a number between brackets. These numbers are used throughout the study to refer to the matrix properties denoted by the indicated expression.

Let us first consider the Kronecker product and the vec-operator. Extensive reviews can for instance also be found in Balestra (1976), Henderson and Searle (1981) and Magnus and Neudecker (1986). In the two latter studies also some historical notes on the origins of the Kronecker product and the vec-operator are given. In what follows the capital A denotes an (m×n) matrix with elements a_{ij} (i = 1,...,m and j = 1,...,n), that is

$$(A.1) \qquad A = \begin{bmatrix} a_{11} & a_{12} & \cdots & a_{1n} \\ a_{21} & a_{22} & & \vdots \\ \vdots & & & \vdots \\ a_{m1} & \cdots & \cdots & a_{mn} \end{bmatrix}$$

Similarly B defines a (p×q) matrix. I_m denotes the (m×m) identity matrix and C, D are two matrices defined to be the appropriate dimensions, so that the matrix additions and multiplications in the following equations are meaningful. The Kronecker product, denoted by ⊗, is defined by

$$
(A.2) \qquad A \otimes B = \begin{bmatrix} a_{11}B & a_{12}B & \cdots & a_{1n}B \\ a_{21}B & a_{22}B & & \vdots \\ \vdots & & & \vdots \\ a_{m1}B & & \cdots & a_{mn}B \end{bmatrix} \qquad (mp \times nq).
$$

Observe that the Kronecker product A ⊗ B is defined for any pair of matrices, independent of their dimensions. The first interest in Kronecker product properties originates from its typical characteristic roots. When A and B are square matrices (that is m=n and p=q) with characteristic roots $\lambda_1, \ldots, \lambda_n$ and $\mu_1, \ldots, \mu_p$, then the characteristic roots of A ⊗ B are given by $\lambda_i \mu_j$ (i = 1,...,m and j = 1,...p). Other important elementary properties of the Kronecker product are

(A.3) $(A \otimes B) \otimes C = A \otimes (B \otimes C)$

(A.4) $(A \otimes B)' = A' \otimes B'$

(A.5) $(A + C) \otimes B = A \otimes B + C \otimes B; \qquad A \otimes (B + D) = A \otimes B + A \otimes D$

(A.6) $(A \otimes B)(C \otimes D) = AC \otimes BD$

(A.7)   if m=n, p=q and A, B non-singular, then $(A \otimes B)^{-1} = A^{-1} \otimes B^{-1}$.

Proofs of these properties can for instance be found in Neudecker (1968). When A and B are replaced by a column and a row vector, then the Kronecker product equals a common matrix product. In formula: when a and b denote two column vectors, then

(A.8) $a \otimes b' = b' \otimes a = ab'$.

By the operator "vec" is meant the columnwise notation of a matrix. Hence denoting by $A_{(i)}$ the ith column of A, we have

(A.9) $vecA = (A'_{(1)}, A'_{(2)}, \ldots, A'_{(n)})'$ $(mn \times 1)$.

Obviously when a and b are two column vectors, then

(A.10) $vec(ab') = b \otimes a$.

Combining vec and $\otimes$ for matrices we obtain the following property

(A.11) $vec(ACB) = (B' \otimes A)vecC$.

When n is equal to p, such that matrix product AB exists, we also have

(A.12) $((vecB)' \otimes I_m)(I_q \otimes vecA) = AB$

and

(A.13) $(I_m \otimes (vecB')')(vecA' \otimes I_q) = AB$.

The connection between the vec operator and the trace operator is given by

(A.14) $tr(AC) = (vecA')'(vecC)$,

where it is assumed that AC is a square matrix. In case of symmetric matrices
of order m, one can define a generalization of the vec-operator, that is often
denoted by vech, which stacks the elements above and including the diagonal of
the matrix, yielding a $(\frac{1}{2}m(m+1) \times 1)$ column vector. (See for instance Henderson
and Searle (1979) and Magnus and Neudecker (1980)). This notation will not be
applied in this study.

In combination with the Kronecker product and the vec operator a useful tool
is the permutation matrix $P_{m,n}$, that is given by

(A.15) $P_{m,n} = [I_n \otimes [I_m]_{(1)}, I_n \otimes [I_m]_{(2)}, \ldots, I_n \otimes [I_m]_{(m)}]$ $(mn \times mn)$

or in words, $P_{m,n}$ is a matrix consisting of $(n \times m)$ blocks each of order $(m \times n)$
and the $(i,j)^{th}$ block has a unit element in the $(j,i)^{th}$ position and zeros
elsewhere. The basic properties of the permutation matrix in connection with $\otimes$

and vec are

(A.16) $P_{m,p}(A \otimes B)P_{q,n} = (B \otimes A)$

and

(A.17) $P_{n,m}(vecA) = vecA'$.

Some elementary properties that follow directly from definition (A.15) are given by

(A.18) $P'_{m,n} = P_{n,m}$

(A.19) $P_{m,1} = P_{1,m} = I_m$

(A.20) $P_{m,n}P_{n,m} = I_{mn}$.

A result that combines $\otimes$, vec and $P_{m,n}$, which is used in this study is given by

(A.21) $((vecI_n)' \otimes I_m)(C \otimes P_{m,n}(A \otimes D))(vecI_n \otimes I_p) = AC'D$

where A is (as usual) an (m×n) matrix, C a square matrix of order n and D denotes an (n×p) matrix.

The following matrix technique, that will be reviewed briefly is the analysis of generalized inverses. Consider a set of equations Ax = y, with an unknown vector x. When A is a square and non-singular matrix, then the (unique) solution for x is known to be given by $x = A^{-1}y$. However, when A is a rectangular or a singular matrix, the inverse A^{-1} does not exist. In this case it is known that the solution for x is not unique and by definition the class of solutions is given by $x = A^-y$. A^- is called the generalized inverse of A and, for instance, in Rao (1973, p. 24) it is shown that

(A.22) A^- is a generalized inverse of A <=> $AA^-A = A$.

In general, when Ax = y is consistent, then its solution is given by

$A^-y + (I_n - A^- A)u$, where A^- is any particular generalized inverse and u is an arbitrary column vector. Note that in case of a square non-singular matrix A, the generalized inverse is unique and equals A^{-1}.

In order to obtain a unique generalized inverse of A it is necessary to impose some further conditions besides (A.22). Various suitable sub-conditions are for instance studied in Bouillon and Odell (1971) and Rao and Mitra (1971). Probably the most widely used unique generalized inverse is defined by Moore (1935) and Penrose (1955). This so-called Moore-Penrose inverse is denoted by A^+, in order to distinguish it from the generalized inverse A^-, and satisfies the properties

(A.23)   (i) $AA^+A = A$ (iii) $(AA^+)' = AA^+$

(ii) $A^+AA^+ = A^+$ (iv) $(A^+A)' = A^+A.$

In some theorems of this study we shall need the generalized inverse of $(\Sigma - \lambda I_k)$ where Σ is a non-singular square, symmetric (covariance) matrix of order k and λ is one of its characteristic roots. It is known that λ is defined by $|\Sigma - \lambda I_k| = 0$, hence $(\Sigma - \lambda I_k)$ itself is a symmetric singular matrix. When its generalized inverse is needed we shall use the Moore-Penrose version, denoted by $(\Sigma - \lambda I_k)^+$. Together with the characteristic root λ, there exists a characteristic (k×1) vector β, say, of Σ such that $\Sigma\beta = \lambda\beta$, or

(A.24) $(\Sigma - \lambda I_k)\beta = 0.$

In the various proofs this relation is observed together with the Moore-Penrose inverse of $(\Sigma - \lambda I_k)$. The question is what (A.23) and (A.24) imply for $(\Sigma - \lambda I_k)^+\beta$. Consider therefore the following result.

Lemma

Let A be an (m×n) matrix and x an (n×1) vector, then

(A.25) $Ax = 0 \iff A'^+x = 0.$

Proof

The generalized Moore-Penrose inverse of A can be represented by $(A'A)^+A'$. Hence A'^+ can be rewritten as $(AA')^+A$, so that from $Ax = 0$ the if-part of (A.25) follows directly. The only-if-part of (A.25) is proved by replacing in the if-part A by A'^+. $\diamond$

From (A.25) it follows that (A.24) yields

(A.26) $(\Sigma - \lambda I_k)^+ \beta = 0$

as $(\Sigma - \lambda I_k)$ is a square, symmetric matrix.

Now we come to the last matrix analysis technique that will be reviewed in
this appendix, viz. the analysis of matrix derivatives. What do we have to
think of when we are talking about differentiating one matrix, with respect to
another matrix? Let us first consider the partial derivatives of vector valued
functions, yielding gradient matrices, as it is known from the traditional
mathematical function analysis. Let ϕ denote a vector function defined on (a
subset of) $\mathbb{R}^k$, with values in $\mathbb{R}^\ell$, which is differentiable on its domain. The
argument of function ϕ is denoted by (k×1) vector x and the function value by
the (ℓ×1) vector $(\phi_1(x),\ldots,\phi_\ell(x))'$. The partial scalar derivatives of ϕ with
respect to x are denoted by

(A.27) $\dfrac{\partial \phi_i(x)}{\partial x_j}$ $i = 1,\ldots,k$ and $j = 1,\ldots,\ell$.

Sorting the partical derivatives in a (k×ℓ) matrix we obtain the gradient
matrix

(A.28) $\nabla(\phi;x) = \dfrac{\partial \phi(x)}{\partial x'} = \left[\dfrac{\partial \phi_i(x)}{\partial x_j} \right]$ (k×ℓ).

One of the first generalizations was to take (a subset of) $\mathbb{R}^{m \times n}$ as domain of
ϕ, with scalar function values. In what follows the argument of the function
will be denoted by A. A straightforward generalization of (A.28) is to order
the partical derivatives in an (m×n) matrix

(A.29) $\dfrac{\partial \phi(A)}{\partial A} = [\dfrac{\partial \phi}{\partial a_{ij}}]$ (m×n) (see for instance Dwyer and MacPhail (1948), Neudecker (1967)). When ϕ is not a scalar valued, but a matrix valued function with function values in $R^{p \times q}$, then notation (A.29) has to be extended. Defining (p×q) matrix B as a function of (m×n) matrix A one possible definition of the matrix derivative of B with respect to A is given by

$$(A.30) \quad \frac{\partial B}{\partial A} = \begin{bmatrix} \dfrac{\partial b_{11}}{\partial A} & \dfrac{\partial b_{12}}{\partial A} & \cdots & \dfrac{\partial b_{1q}}{\partial A} \\[2ex] \dfrac{\partial b_{21}}{\partial A} & \dfrac{\partial b_{22}}{\partial A} & & \vdots \\[2ex] \vdots & & & \vdots \\[2ex] \dfrac{\partial b_{p1}}{\partial A} & \cdots & \cdots & \dfrac{\partial b_{pq}}{\partial A} \end{bmatrix} \qquad (mp \times nq)$$

with block $\dfrac{\partial b_{ij}}{\partial A}$ similar to (A.29) defined by

$$(A.31) \quad \frac{\partial b_{ij}}{\partial A} = \begin{bmatrix} \dfrac{\partial b_{ij}}{\partial a_{11}} & \dfrac{\partial b_{ij}}{\partial a_{12}} & \cdots & \dfrac{\partial b_{ij}}{\partial a_{1n}} \\[2ex] \dfrac{\partial b_{ij}}{\partial a_{21}} & \dfrac{\partial b_{ij}}{\partial a_{22}} & & \vdots \\[2ex] \vdots & & & \vdots \\[2ex] \dfrac{\partial b_{ij}}{\partial a_{m1}} & \cdots & \cdots & \dfrac{\partial b_{ij}}{\partial a_{mn}} \end{bmatrix}.$$

This definition presents one of the various ways, to display the mnpq partial derivatives in a 2-dimensional array. In this study, foregoing definition of matrix differentiation will be used to derive the solution of minimization problems. (For applications of matrix derivatives see also Dwyer (1967), Tracy and Dwyer (1969), MacRae (1974) and Balestra (1976)). However, this definition does not form a natural generalization of the notation of a gradient matrix of a vector function to a gradient matrix of a matrix function. For instance the matrix of partial derivatives of the identity function (that is B equals A) is according to definition (A.30)-(A.31) given by

(A.32) $\frac{\partial A}{\partial A} = (vecI_m)(vecI_n)'$

and does not equal I_{mn} as one should expect for the identity function. Neither does the practical chainrule, as defined for vector function differentiation, apply to this definition of matrix differentiation. More critical remarks on this definition can be found in Magnus and Neudecker (1985) and Pollock (1985).

Neudecker (1969) proposed to vectorize the matrices, before taking derivatives, yielding the following definition of (vectorized) matrix derivatives

(A.33) $\frac{\partial (vecB)}{\partial (vecA)'} = [\frac{\partial b_{11}}{\partial (vecA)}, \frac{\partial b_{21}}{\partial (vecA)}, \cdots, \frac{\partial b_{p1}}{\partial (vecA)}, \frac{\partial b_{12}}{\partial (vecA)}, \cdots, \frac{\partial b_{pq}}{\partial (vecA)}]'$

$(pq \times mn)$

with for $i = 1, \ldots, p$ and $j = 1, \ldots, q$

(A.34) $\frac{\partial b_{ij}}{\partial (vecA)} = (\frac{\partial b_{ij}}{\partial a_{11}}, \frac{\partial b_{ij}}{\partial a_{21}}, \cdots, \frac{\partial b_{ij}}{\partial a_{m1}}, \frac{\partial b_{ij}}{\partial a_{12}}, \cdots, \frac{\partial b_{ij}}{\partial a_{mn}})'$ $(mn \times 1)$.

Note that in fact matrix derivative definition (A.30)-(A.31) is applied to the $(1 \times mn)$ and $(pq \times 1)$ "matrices" $(vecA)'$ and $vecB$. Magnus and Neudecker (1985) and Pollock (1985) recommended this way of collecting the partial derivatives as it does not lead to serious disjunctions of the familiar gradient matrix, as is the case in other types of matrix derivative definitions. For instance, the chainrule applies to definition (A.33)-(A.34). That is, when matrix C is observed as a function of matrix B, then

(A.35) $\quad \dfrac{\partial(vecC)}{\partial(vecA)^\intercal} = \dfrac{\partial(vecC)}{\partial(vecB)^\intercal} \dfrac{\partial(vecB)}{\partial(vecA)^\intercal}.$

Furthermore does the identity function yield

(A.36) $\quad \dfrac{\partial(vecA)}{\partial(vecA)^\intercal} = I_{mn}.$

Because of the similarities between the gradient matrix for vector derivatives and the definition of vectorized matrices we define similar to (A.28) a gradient matrix for vector functions by

(A.37) $\quad \nabla(B;A) = \dfrac{\partial(vecB)}{\partial(vecA)^\intercal}.$

The matrix derivatives that are needed in this study in connection with the application of the delta method (see chapter II for this statistical device) are given in the $\nabla(.;.)$ notation. However, in the derivations of expressions for these gradient matrices also the matrix derivative notation as defined in (A.30)-(A.31) is used. It may be observed that one may get more economic expressions if A and/or B are square and symmetric. As the use of the symmetry would entail additional notations (the use of vech instead of vec, see also McCulloch (1982)) and does not yield any essential advantage in the application of the delta method, this way of collecting the partial matrix derivatives will not be pursued in this appendix.

In what follows some elementary matrix derivative results are given. Proofs of these results may be found in one or more of the aforementioned references on matrix differentiation. Again it is assumed that the matrices A and B have dimensions (mxn) and (pxq) respectively. C and D denote matrices with dimensions such that a given matrix operation applied to them is well defined. The simple transpose operation yields the derivative

(A.38) $\quad \dfrac{\partial A'}{\partial A} = P_{m,n}$

and as a result we have

(A.39) $\quad \dfrac{\partial A}{\partial A'} = (\dfrac{\partial A'}{\partial A})' = P_{n,m}.$

Sum and product rules for matrix derivatives are given by

(A.40) $\quad \dfrac{\partial(C+D)}{\partial A} = \dfrac{\partial C}{\partial A} + \dfrac{\partial D}{\partial A}$

and

(A.41) $\quad \dfrac{\partial(CD)}{\partial A} = \dfrac{\partial C}{\partial A}(D \otimes I_n) + (C \otimes I_m)\dfrac{\partial D}{\partial A}.$

Using (A.32), (A.38) and productrule (A.41) it can easily be shown that when C and D are two matrices that are not functionally dependent on A

(A.42) $\quad \dfrac{\partial(CAD)}{\partial A} = (\text{vec}C')(\text{vec}D)'$

and

(A.43) $\quad \dfrac{\partial(CA'D)}{\partial A} = P_{m,p}(D \otimes C) = (C \otimes D)P_{q,n}$

where p denotes the number of rows of C and q denotes the number of columns of D. Somewhat less easy to prove is the following expression for matrix quadratic forms with constant matrix C

(A.44) $\quad \dfrac{\partial(A'CA)}{\partial A} = P_{m,n}(CA \otimes I_n) + (\text{vec}CA)(\text{vec}I_n)'.$

One can also observe the derivatives of values, obtained after applying a matrix operator to B. For instance observing the vec operator one can derive

(A.45) $\quad \dfrac{\partial(\text{vec}B)}{\partial A} = (I_q \otimes \dfrac{\partial B}{\partial A})(\text{vec}I_q \otimes I_n)$

and

(A.46) $\quad \dfrac{\partial B}{\partial(\text{vec}A)'} = (I_p \otimes (\text{vec}I_m)')(\dfrac{\partial B}{\partial A} \otimes I_m).$

A special case of (A.46) is obtained by taking B = A. Application of (A.32) yields

(A.47) $\dfrac{\partial A}{\partial (vecA)'} = ((vecI_n)' \otimes I_m).$

Another simple matrix operator that can be considered is the trace operator. The trace of a square matrix is defined as the sum of its diagonal elements. The matrix derivative of trB, with B a square (p×p) matrix that is a function of A, is given by

(A.48) $\dfrac{\partial\ trB}{\partial A} = ((vecI_p)' \otimes I_m)(I_p \otimes \dfrac{\partial B}{\partial A})(vecI_p \otimes I_n).$

For some special forms of B, as considered in (A.42), (A.43) and (A.44) one can prove

(A.49) $\dfrac{\partial\ tr(CAD)}{\partial A} = C'D',$

(A.50) $\dfrac{\partial\ tr(CA'D)}{\partial A} = DC$

and

(A.51) $\dfrac{\partial\ tr(A'CA)}{\partial A} = (C + C')A.$

The final matrix operation observed in the context of matrix differentiation is the matrix inversion. For a general matrix B, the derivative of its inverse is given by

(A.52) $\dfrac{\partial B^{-1}}{\partial A} = -(B^{-1} \otimes I_m) \dfrac{\partial B}{\partial A} (B^{-1} \otimes I_n),$

provided that B is non-singular. As a special case one obtains by applying (A.32)

(A.53) $\dfrac{\partial A^{-1}}{\partial A} = -(vec(A^{-1})')(vec(A^{-1})').$

References

Afriat, S.N. (1957), "Orthogonal and oblique projectors and the characteristics of pairs of vector spaces", Proceedings of the Cambridge Philosophical Society 53, pp. 800-816.

Aitchison, J. and S.D. Silvey (1958), "Maximum-likelihood estimation of parameters subject to restraints", Annals of Mathematical Statistics 29, pp. 813-828.

Aitchison, J. and S.D. Silvey (1960), "Maximum-likelihood estimation procedures and associated tests of significance", Journal of the Royal Statistical Society, Series B 22, pp. 154-171.

Akaike, H. (1981), "Likelihood of a model and information criteria", Journal of Econometrics (Annals of Applied Econometrics) 16, pp. 3-14.

Amemiya, T. (1980), "Selection of regressors", International Economic Review 21, pp. 331-354.

Anderson, T.W. (1951), "The asymptotic distribution of certain characteristic roots and vectors", Proceedings of the Second Berkeley Symposium on Mathematical Statistics and Probability, pp. 103-130, University of California Press, Berkeley.

Anderson, T.W. (1958), An Introduction to Multivariate Statistical Analysis, Wiley, New York.

Anderson, T.W. (1963), "Asymptotic theory for principal component analysis", Annals of Mathematical Statistics 34, pp. 122-148.

Anderson, T.W. (1984), The 1982 Wald Memorial Lectures, "Estimating linear statistical relationships", The Annals of Statistics 12, pp. 1-45.

Atkinson, A.C. (1981), "Likelihood ratios, posterior odds and informattion criteria", Journal of Econometrics (Annals of Applied Econometrics) 16, pp. 15-20.

Bailar, B.A. and J.C. Bailar III (1978), "Comparison of two procedures for imputing missing survey values", American Statistical Association, Proceedings of the Section on Survey Research Methods, pp. 462-467.

Balestra, P. (1976), La Derivation Matricielle, Collection de l'Institute de Mathématiques Economiques, no. 12, Sirey, Paris.

Bartlett, M.S. (1950), "Tests of significance in factor analysis", British Journal of Psychology; Statistical Section 3, pp. 77-85.

Bartlett, M.S. (1951), "A further note on tests of significance in factor analysis", British Journal of Psychology; Statistical Section 4, pp. 1-2.

Basmann, R.L. (1957), "A generalized classical method of linear estimation of coefficients in a structural equation", Econometrica 27, pp. 72-81.

Basmann, R.L. (1959), "The computation of generalized classical estimates of coefficients in a structural equation", Econometrica 27, pp. 72-81.

Basmann, R.L. (1960), "On the asymptotic distribution of generalized linear
    estimators", Econometrica 28, pp. 97-108.

Bellman, R. (1970), Introduction to Matrix Analysis, McGraw-Hill, New York.

Bentler, P.M. (1983), "Some contributions to efficient statistics in
    structural models: specification and estimation of moment structures",
    Psychometrika 48, pp. 493-517.

Bentler, P.M. and T.K. Dijkstra (1983), "Efficient estimation via
    linearization in structural models", in: Multivariate Analysis VI, P.R.
    Krishnaiah (ed.), North-Holland, Amsterdam.

Benzecri, J.P. et al. (1973), L'Analyse des Données, Vol. 1: La Taxinomie,
    Vol. 2: Correspondences, Dunod, Paris.

Binkley, J.K. (1981), "The effect of variable correlation on the efficiency of
    seemingly unrelated regression in a two equation model", Journal of the
    American Statistical Association 77, pp. 890-895.

Bishop, Y.M.M., S.E. Fienberg and P.W. Holland (1975), Discrete Multivariate
    Analysis: Theory and Practice, MIT Press, Cambridge Massachusetts.

Bouillon, T.L. and P.L. Odell (1971), Generalized Inverse Matrices, Wiley, New
    York.

Box, G.E.P. (1979), "Some problems of statistics and everyday life", Journal
    of the American Statistical Association 74, pp. 1-4.

Breusch, T.S. and A.R. Pagan (1980), "The Lagrange multiplier test and its
    applications to model specification in econometrics", Review of Economic
    Studies 47, pp. 239-253.

Browne, M.W. (1974), "Generalized least squares estimators in the analysis of
    covariance structures", South African Statistical Journal 8, pp. 1-24.

Browne, M.W. (1984), "Asymptotically distribution-free methods for the
    analysis of covariance structures", British Journal of Mathematical and
    Statistical Psychology 37, pp. 62-83.

Byron, R.P. (1970), "The restricted Aitken estimation of sets demand
    relations", Econometrica 38, pp. 816-830.

Byron, R.P. (1972), "Testing for misspecification in econometric systems
    using full information", International Economic Review 13, pp. 745-756.

Chamberlain, G. (1982), "Multivariate regression models for panel data",
    Journal of Econometrics (Annals of Applied Econometrics), 18, pp. 5-46.

Chernoff, H. (1956), "Large sample theory: parametric case", Annals of
    Mathematical Statistics 27, pp. 1-22.

Cramer, J.S. (1964), "Efficient grouping, regression and correlation in Engel
    curve analysis", Journal of the American Statistical Association 59, pp.
    233-250.

Darroch, J.N. (1965), "An optimal property of principal components", _Annals of Mathematical Statistics_ **36**, pp. 1579-1582.

De Leeuw, J. (1983), "Models and Methods for the analysis of correlation coefficients", _Journal of Econometrics_ (Annals of Applied Econometrics) **22**, pp. 113-138.

De Leeuw, J. (1984), "Models of data", _Kwantitatieve Methoden_ **13**, pp.17-30.

De Leeuw, J. (1986), "Model selection in multinomial experiments", Internatnal Report, Department of Data Theory, University of Leyden.

Dent, W.T. (1980), "On restricted estimation in linear models", _Journal of Econometrics_ **12**, pp. 49-58.

Deville, J.C. and E. Malinvaud (1983), "Data analysis in official socio-economic statistics", Journal of the Royal Statistical Society, Series B, pp. 335-361.

Dhrymes, P.J. (1978), _Introductory Econometrics_, Springer Verlag, New York.

Dhrymes, P.J. (1980), _Econometrics; Statistical Foundations and Applications_, Springer Verlag, New York.

Dijkstra, T.K. (1984), (ed.) _Misspecification Analysis_, Springer-Verlag, Berlin.

Durbin, J. (1953), "A note on regression when there is extraneous information about one of the coefficients", _Journal of the American Statistical Association_ **48**, pp. 799-808.

Dwyer, P.S. (1967), "Some applications of matrix derivatives in multivariate analysis", _Journal of the American Statistical Association_ **62**, pp. 607-625.

Dwyer, P.S. and M.S. MacPhail (1948), "Symbolic matrix derivatives", _Annals of Mathematical Statistics_ **19**, pp. 517-534.

Feller, W. (1966), _An Introduction to Probability Theory and Its Applications_, Vol. II, Wiley, New York.

Fisher, R.A. (1915), "Frequency distribution of the values of the correlation coefficient in samples from an infinitely large population", _Biometrika_ **10**, pp. 507-521.

Fisher, R.A. (1921), "On the 'probable error' of a coefficient of correlation deduced from a small sample", _Metron_ 1, pp. 1-32.

Fisher, R.A. (1928), "The general sampling distribution of the multiple correlation coefficient", _Proceedings of the Royal Society of London, Series A_ **121**, pp. 654-673.

Fisher, R.A. (1925), _Statistical Methods for Research Workers_, Oliver and Boyd, Edinburgh.

Fisher, R.A. (1935), _The Design of Experiments_, Oliver and Boyd, Edinburgh.

Ford, B.L. (1980), "An overview of hot-deck procedures", American Statistical Association, Proceedings of the Section on Survey Research Methods, pp. 87-125.

Friedrichs, K.O. (1965), "Perturbation of spectra in Hilbert space", American Mathematical Society, Providence, Rhode Island.

Gallant, A.R. and T.M. Gerig, (1980), "Computations for constrained linear models", Journal of Econometrics 12, pp. 59-84.

Gandolfo, G. (1981), Qualitative Analysis and Econometric Estimation of Continuous Time Dynamic Models, North-Holland, Amsterdam.

Gantmacher, F.R. (1959), The Theory of Matrices, Vol. I and II, Chelsea, New York.

Gifi, A. (1984), Nonlinear Multivariate Analysis, DSWO-press, Leyden.

Girshick, M.A. (1939), "On the sampling theory of roots of determinantal equations", Annals of Mathematical Statistics 10, pp. 203-224.

Glahn, H.R. (1969), "Some relationships derived from canonical correlation theory", Econometrica 37, pp. 252-256.

Glynn, W.J. and R.J. Muirhead (1978), "Inference in canonical correlation analysis", Journal of Multivariate Analysis 8, pp. 468-478.

Goedhart, T., V. Halberstadt, A. Kapteyn and B.M.S. van Praag (1977), "The poverty line: concept and measurement", Journal of Human Resources 12, pp. 503-520.

Goldberger, A.S. (1964), Econometric Theory, Wiley, New York.

Golub, G.H. and C.F. Van Loan (1983), Matrix Computations, North Oxford Academic, Oxford.

Haavelmo, T. (1943), "The statistical implications of a system of simultaneous equations", Econometrica 11, pp. 1-12.

Haavelmo, T. (1944), "The probability approach in econometrics", Econometrica 12, Suppl., pp. 1-118.

Haavelmo, T. (1947), "Methods of measuring the marginal propensity to consume", Journal of the American Statistical Association 42, pp. 105-122.

Haitovsky, Y. (1968), "Missing data in regression analysis", Journal of the Royal Statistical Society, Series B 32, pp. 67-82.

Hansen, L.P. (1982), "Large sample properties of generalized method of moments estimators", Econometrica 50, pp. 1029-1054.

Harman, H.H. (1967), Modern Factor Analysis, University of Chicago Press, Chicago.

Hausman, J.A. (1978), "Specification tests in econometrics", Econometrica 46, pp. 1251-1272.

Henderson, H.V. and S.R. Searle (1979), "Vec and vech operators for matrices, with some uses in Jacobians and multivariate statistics", The Canadian Journal of Statistics 7, pp. 65-81.

Henderson, H.V. and S.R. Searle (1981), "The vec-permutation matrix, the vec operator and Kronecker products: a review", Linear and Multilinear Algebra 9, pp. 271-288.

Homan, M.E., A.J.M. Hagenaars and B.M.S. van Praag (1987), "A comparison of six methods to estimate the monetary value of home production", Discussion paper, Econometric Institute, Erasmus University, Rotterdam.

Hooper, J.W. (1959), "Simultaneous equations and canonical correlation theory", Econometrica 27, pp. 245-256.

Hooper, J.W. (1962), "Partial trace correlations", Econometrica 30, pp. 324-331.

Hotelling, H. (1933), "Analysis of a complex of statistical variables into principle components", Journal of Educational Psychology 26, pp. 417-441, 498-520.

Hotelling, H. (1935), "The most predictable criterion", Journal of Educational Psychology 26, pp. 139-142.

Hotelling, H. (1936), "Relations between two sets of variates", Biometrika 28, pp. 321-377.

Hsu, P.L. (1941), "On the limiting distribution of canonical correlations", Biometrika 32, pp. 38-45.

I.B.M. Programmers Manual, System/360 Scientific Subroutine Package, Version III (1970), I.B.M. Corporation, Technical Publications Department, New York.

James, A.T. (1964), "Distributions of matrix variates and latent roots derived from normal samples", Annals of Mathematical Statistics 35, pp. 475-501.

Johansson, J.K. (1981), "An extension of Wollenberg's redundancy analysis", Psychometrika 46, pp. 93-103.

Judge, G.G., R.C. Hill, W.E. Griffiths, H. Lütkepohl and T.C. Lee (1982), Introduction to the Theory and Practice of Econometrics, Wiley, New York.

Kakwani, N.C. (1967), "The unbiasedness of Zellner's seemingly unrelated regression equations estimate", Journal of the American Statistical Association 62, pp. 141-142.

Kalman, R.E. (1982,a), "Identification from real data", Current Developments in the Interface: Economics, Econometrics, Mathematics, M. Hazewinkel and A.H.G. Rinnooy Kan (eds.), pp. 161-196, Reidel.

Kalman, R.E. (1982,b), "System identification from noisy data", Dynamical Systems II, A.R. Bednarek and L. Lesari (eds.), pp. 135-164, Academic Press, New York.

Kalman, R.E. (1983), "Identifiability and modeling in econometrics", Developments in Statistics, P.R. Krishnaiah (ed.), pp. 97-136, Academic Press, New York

Kato, T. (1955), "Quadratic forms in Hilbert spaces and asymptotic perturbation series", Berkeley, University of California.

Kato, T. (1966), Perturbation Thoery for Linear Operators, Springer, Berlin.

Kelker, D. (1970), "Distribution theory of spherical distributions and a location-scale parameter generalization", Sankhya, Series A 32, pp. 419-430.

Keller, W.J. and T.J. Wansbeek (1983), "Multivariate methods for quantitative and qualitative data", Journal of Econometrics (Annals of Applied Econometrics) 22, pp. 91-111.

Kendall, M.G. (1951), "Regression, structure and functional relationship. I", Biometrika 38, pp. 11-25.

Kendall, M.G. (1952), "Regression, structure and functional relationship. II", Biometrika 39, pp. 96-108.

Kendall, M.G. and A. Stuart (1961), The Advanced Theory of Statistics, vol. II, Griffin, London.

Kettenring, J.R. (1971), "Canonical analysis of several sets of variables", Biometrika 58, pp. 433-451.

Khatri, C.G. and K.C.S. Pillai (1969), "Distributions of vectors corresponding to the largest roots of three matrices", Multivariate Analysis II, P.R. Krishnaiah (ed.), pp. 219-240, Academic Press, New York.

Kiviet, J.F., (1987), Testing Linear Econometric Models, Dissertation, University of Amsterdam.

Koerts, J. and A.P.J. Abrahamse (1970), "The correlation coefficient in the general linear model", European Economic Review 1, pp. 401-427.

Laughton, M.A. (1964), "Sensitivity in dynamical system analysis", Journal of Electronics and Control 17, pp. 577-591.

Lawley, D.N. (1940), "The estimation of factor loadings by the method of maximum likelihood", Proceedings of the Royal Society of Edinburgh 60, pp. 64-82.

Lawley, D.N. (1942), "Further investigation in factor estimation", Proceedings of the Royal Society of Edinburgh 61, pp. 176-185.

Lawley, D.N. (1943), "The application of the maximum likelihood method to factor analysis", British Journal of Psychology 33, pp. 172-175.

Lawley, D.N. (1956), "Tests of significance for the latent roots of covariance and correlation matrices", Biometrika 43, pp. 128-136.

Lawley, D.N. (1959), "Tests of significance in canonical analysis", *Biometrika* **46**, pp. 59-66.

Lawley, D.N. and A.E. Maxwell (1963), *Factor Analysis as a Statistical Method*, Butterworth's, London.

Lebart, L., A. Morineau and N. Tabard (1977), *Techniques de la Description Statistique*, Dunod, Paris.

Luenberger, D.G. (1969), *Optimization by Vector Space Methods*, Wiley, New York.

MacRae, E.C. (1974), "Matrix derivatives with an application to an adaptive linear decision model", *The Annals of Statistics* **2**, pp. 337-346.

Madansky, A. (1964), "On the efficiency of three-stage least squares estimation", *Econometrica* **32**, pp. 51-56.

Maddala, G.S. (1979), *Econometrics*, McGraw-Hill, New York.

Magnus, J.R. and H. Neudecker (1979), "The commutation matrix: some properties and applications", *The Annals of Statistics* **7**, pp. 381-394.

Magnus, J.R. and H. Neudecker (1980), "The elimination matrix: some lemmas and applications", *SIAM Journal on Algebraic and Discrete Methods* **1**, pp. 422-449.

Magnus, J.R. and H. Neudecker (1985), "Matrix differential calculus with applications to simple, Hadamard, and Kronecker products", *Journal of Mathematical Psychology* **29**, pp. 474-492.

Magnus, J.R. and H. Neudecker (1986), "Symmetry, 0-1 matrices and Jacobians: a review", *Econometric Theory* **2**, pp. 157-190.

Mak, T.K. (1981), "Large sample results in the estimation of a linear transformation", *Biometrika* **68**, pp. 323-325.

Malinvaud, E. (1970), *Statistical Methods of Econometrics*, North-Holland, Amsterdam.

Mardia, K.V., J.T. Kent and J.M. Bibby (1979), *Multivariate Analysis*, Academic Press, New York.

McGullagh, P. and J.A. Nelder (1983), *Generalized Linear Models*, Chapman and Hall, London.

McCulloch, C.E. (1982), "Symmetric matrix derivatives with applications", *Journal of the American Statistical Association* **77**, pp. 679-682.

McElroy, M.B. (1977), "Goodness of fit for seemingly unrelated regressions", *Journal of Econometrics* **6**, pp. 381-387.

McKeon, J.J. (1965), "Canonical analysis: Some relations between canonical correlation, factor analysis, discriminant function analysis, and scaling theory", *Pscyhometrik Monographs* **13**, University of Chicago Press, Chicago.

Mood, M.A., F.A. Graybill and D.C. Boes (1974), Introduction to the Theory of Statistics, third edition, McGraw-Hill, New York.

Moore, E.H. (1935), General Analysis, American Philosophical Society, Philadelphia.

Morrison, D.F. (1978), Multivariate Statistical Methods, McGraw-Hill, New York.

Muirhead, R.J. (1980), "The effects of elliptical distributions on some standard procedures involving correlation coefficients: A review", Multivariate Statistical Analysis, R.P. Gupta (ed.), pp. 143-159, North-Holland, Amsterdam.

Muirhead, R.J. (1982), Aspects of Multivariate Statistical Theory, Wiley, New York.

Muirhead, R.J. and C.M. Waternaux (1980), "Asymptotic distributions in canonical correlation analysis and other multivariate procedures for nonnormal populations", Biometrika 67, pp. 31-43.

Muller, K.E. (1981), "Relationships between redundancy analysis, canonical correlation, and multivariate regression", Psychometrika 46, pp. 139-142.

Neudecker, H. (1967), "On matrix procedures for optimizing differentiable scalar functions of matrices", Statistica Neerlandica 21, pp. 101-106.

Neudecker, H. (1968), "The Kronecker product and some of its applications in econometrics", Statistica Neerlandica 22, pp. 69-82.

Neudecker, H. (1969), "Some theorems on matrix differentiation with special reference to Kronecker matrix products", Journal of the American Statistical Association 64, pp. 953-963.

Neudecker, H. and T.J. Wansbeek (1983), "Some results on commutation matrices, with statistical applications", The Canadian Journal of Statistics 11, pp. 221-231.

Okamoto, M. (1969), "Optimality of principal components", Multivariate Analysis II, P.R. Krishnaiah (ed.), pp. 673-685, Academic Press, New York.

Okamoto, M. and M. Kanazawa (1968), "Minimization of eigen values of a matrix and optimality of principal components", Annals of Mathematical Statistics 39, pp. 859-863.

Pearson, K. (1894), "Contributions to the mathematical theory of evolution", Philosophical Transactions of the Royal Society of London, Series A 185, pp. 71-78.

Penrose, R. (1955), "A generalized inverse for matrices", Proceedings of the Cambridge Philosophical Society, pp. 406-413.

Pesaran, M.H. (1974), "On the general problem of model selection", Review of Economic Studies 41, pp. 153-171.

Phillips, P.C.B. (1982), "A simple proof of the latent root sensitivity formula", Economics Letters 9, pp. 57-59.

Pierce, D.A. (1982), "The asymptotic effect of substituting estimators for parameters in certain types of statistics", The Annals of Statistics 10, pp. 475-478.

Plackett, R.L. (1972), "Studies in the history of probability and statistics XXIX. The discovery of the method of least squares", Biometrika 59, pp. 239-251.

Pollock, D.S.G. (1979), The Algebra of Econometrics, Wiley, New York.

Pollock, D.S.G. (1985), "Tensor products and matrix differential calculus", Linear Algebra and Its Applications 67, pp. 169-193.

Pratt, J. (1959), "On a general concept of 'inprobability'", Annals of Mathematical Statistics 30, pp. 549-558.

Ramsey, J.B. (1969), "Tests for specification errors in classical linear least squares regression analysis", Journal of the Royal Statistical Society, Series B 31, pp. 350-371.

Randles, R.H. (1982), "On the asymptotic normality of statistics with estimated parameters", The Annals of Statistics 10, pp. 462-474.

Rao, C.R. (1964), "The use and interpretation of principal component analysis in applied research", Sankhyā, Series A 26, pp. 329-358.

Rao, C.R. (1973), Linear Statistical Inference and Its Applications, second edition, Wiley, New York.

Rao, C.R. and S.K. Mitra (1971), Generalized Inverse of Matrices and its Applications, Wiley, New York.

Revankar, M.S. (1974), "Some finite sample results in the context of two seemingly unrelated regression equations", Journal of the American Statistical Association 69, pp. 187-190.

Robertson, C.A. (1974), "Large sample theory for the linear structural relation", Biometrika 61, pp. 353-359.

Rothenberg, T.J. and C.T. Leenders (1964), "Efficient estimation of simultaneous equation systems", Econometrica 32, pp. 57-76.

Sande, I.G. (1979), "A personal view of hot-deck imputation procedures", Survey Methodology 5, pp. 238-258.

Sargan, J.D. (1964), "Three-stage least squares and full information maximum likelihood estimates", Econometrica 32, pp. 77-81.

Schmidt, P. (1977), "Estimation of seemingly unrelated regressions with unequal numbers of observations", Journal of Econometrics 5, pp. 365-378.

Serfling, R.J. (1980), Approximation Theorems of Mathematical Statistics, Wiley, New York.

Shilov, G.E. (1977), Linear Algebra, Dover, New York.

Shu, P.L. (1949), "The limiting distribution of functions of sample means and application to testing hypotheses", Proceedings First Berkeley Symposium in Mathematics, Statistics and Probability, pp. 359-402, University of California Press, Berkeley.

Silvey, S.D. (1959), "The Lagrangian multiplier test", Annals of Mathematical Statistics 30, pp. 389-407.

Simon, J.L. (1970), "The concept of causality in economics", Kyklos 23, pp. 226-254.

Soong, T.T. (1969), "An extension of the moment method in statistical estimation", SIAM Journal of Applied Mathematics 17, pp. 560-568.

Spearman, C. (1904), "General intelligence, objectively determined and measured", American Journal of Psychology 15, pp. 201-293.

Steiger, J.H. and A.R. Hakstian (1982), "The asymptotic distribution of elements of a correlation matrix: Theory and application", British Journal of Mathematical and Statistical Psychology 35, pp. 208-215.

Sugiyama, T. (1966), "On the distribution of largest latent root and the corresponding latent vector for principal component analysis", Annals of Mathematical Statistics 37, pp. 995-1001.

Tayler, E.T. (1981), "Asymptotic inference for eigen vectors", The Annals of Statistics 9, pp. 725-736.

Theil, H. (1953), "Estimation and simultaneous correlation in complete equation systems", Central Plan Bureau, The Hague.

Theil, H. (1958), Economic Forecasts and Policy, North-Holland, Amsterdam.

Theil, H. (1971), Principles of Econometrics, Wiley, New York.

Theil, H. and A.S. Goldberger (1961), "On pure and mixed statistical estimation in economics", International Economic Review 2, pp. 65-78.

Thurstone, L.L. (1945), Multiple Factor Analysis, University of Chicago Press, Chicago.

Tracy, D.S. and P.S. Dwyer (1969), "Multivariate maxima and minima with matrix derivatives", Journal of the American Statistical Association 64, pp. 1576-1594.

Tukey, J.W. (1977), Exploratory Data Analysis, Addison-Wesley, London.

Tukey, J.W. (1980), "We need both exploratory and confirmatory", American Statistician 34, pp. 23-25.

Tyler, E.D. (1982), "On the optimality of the simultaneous redundancy transformations", Psychometrika 47, pp. 77-86.

Van de Geer, J.P. (1971), Introduction to Multivariate Analysis for the Social Sciences, Freeman, San Francisco.

Van den Wollenberg, A.L. (1977), "Redundancy analysis. An alternative for canonical analysis", Psychometrika 42, pp. 207-219.

Van Praag, B.M.S. (1971), "The welfare function of income in Belgium; an empirical investigation", European Economic Review 2, pp. 337-369.

Van Praag, B.M.S. (1978), "The multivariate approach in linear regression theory", Compstat 78; Proceedings in Computational Statistics, L.C.A. Korsten and J. Hermans (eds.), Physica-Verlage, Vienna.

Van Praag, B.M.S. (1980), "Model free regression", Report 80.05, Center for Research in Public Economics, Leyden University.

Van Praag, B.M.S. (1981), "Model free regression", Economics Letters 7, pp. 139-144.

Van Praag, B.M.S. (1982), "The Population-Sample Decomposition with an application to minimum distance estimators", Report 82.18, Center for Research in Public Economics, Leyden University.

Van Praag, B.M.S., J. De Leeuw and T. Kloek (1986), "The population-sample decomposition approach to multivariate estimation methods", Applied Stochastic Models and Data Analysis, 2, pp. 99-119.

Van Praag, B.M.S., T.K. Dijkstra and J. Van Velzen (1985), "Least-squares theory based on general distributional assumptions with an application to the incomplete observations problem", Psychometrika 50, pp. 25-36.

Van Praag, B.M.S., T. Goedhart and A. Kapteyn (1980), "The poverty line - A pilot survey in Europe", The Review of Economics and Statistics 62, pp. 461-465.

Van Praag, B.M.S. and J.T.A. Koster (1984), "Specification in simultaneous linear equations models: The relation between a priori specifications and resulting estimators", Misspecification Analysis, T.K. Dijkstra (ed.), pp. 71-84, Springer Verlag, Berlin.

Van Praag, B.M.S. and A.M. Wesselman (1984), "The hot-deck method: an analytical and empirical evaluation", Computational Statistics Quarterly 1, pp. 205-231.

Van Praag, B.M.S. and A.M. Wesselman (1986), "Elliptical multivariate analysis", discussion paper, Econometric Institute, Erasmus University Rotterdam.

Wesselman, A.M. and B.M.S. Van Praag (1984), "Asymptotic distributions of estimators derived from a selective sample", report 84.21, Center for Research in Public Economics, Leyden University.

Wesselman, A.M. and B.M.S. Van Praag (1987), "Elliptical regression operationalized", Economics Letters, forthcoming.

White, H. (1980), "Using least squares to approximate unknown regression functions", International Economic Review 21, pp. 149-170.

White, H. (1981), "Consequences and detection of misspecified nonlinear regression models", Journal of the American Statistical Association 76, pp. 419-433.

White, H. (1982), "Maximum likelihood estimation of misspecified models", Econometrica 50, pp. 1-25.

White, H. (1984), Asymptotic Theory for Econometricians, Academic Press Inc., Orlando, Florida, U.S.A.

Wilks, S.S. (1962), Mathematical Statistics, Wiley, New York.

Wishart, J. (1931), "The mean and second moment coefficient of the multiple correlation coefficient in samples from a normal population", Biometrika 20, pp. 32-52.

Wold, H. (1964), Econometric Model Building: Essays on the Causal Chain Approach, North-Holland, Amsterdam.

Yohai, V.J. and M.S. Garcia Ben (1980), "Canonical variables as optimal predictors", The Annals of Statistics 8, pp. 865-869.

Zellner, A. (1962), "An efficient method of estimating seemingly unrelated regressions and tests of aggregation bias", Journal of the American Statistical Association 57, pp. 348-368.

Zellner, A. and D.S. Huang (1962), "Further properties of efficient estimator for seemingly unrelated regression", International Economic Review 3, pp. 300-313.

Zellner, A. and H. Theil (1962), "Three stage least squares: simultaneous estimation of simultaneous equations", Econometrica 30, pp. 54-78.

Author Index

Subject Index